Water Markets

Water Markets
A Global Assessment

Edited by

Sarah Ann Wheeler

Professor of Water Economics, School of Economics and Public Policy, University of Adelaide, Australia

Cheltenham, UK • Northampton, MA, USA

Published by
Edward Elgar Publishing Limited
The Lypiatts
15 Lansdown Road
Cheltenham
Glos GL50 2JA
UK

Edward Elgar Publishing, Inc.
William Pratt House
9 Dewey Court
Northampton
Massachusetts 01060
USA

A catalogue record for this book
is available from the British Library

Library of Congress Control Number: 2021938804

This book is available electronically in the **Elgar**online
Economics subject collection
http://dx.doi.org/10.4337/9781788976930

ISBN 978 1 78897 692 3 (cased)
ISBN 978 1 78897 693 0 (eBook)

Printed and bound by CPI Group (UK) Ltd, Croydon, CR0 4YY

Contents

Figures

Tables

Boxes

Contributors

Keshab Dhoj Adhikari, Joint Secretary, Ministry of Energy, Nepal.

Rida Afzal, Department of Agribusiness and Applied Economics, MNS-University of Agriculture, Pakistan.

Muhammad Ashfaq, Institute of Agriculture and Resource Economics, University of Agriculture, Pakistan.

Irfan Ahmad Baig, Department of Agribusiness and Applied Economics, MNS-University of Agriculture, Pakistan.

Rosalind H. Bark, School of Environmental Sciences, University of East Anglia, United Kingdom.

Pilar Barria, Facultad de Ciencias Forestales y de la Conservación de la Naturaleza, Universidad de Chile, Chile.

Madhav Belbase, Ministry of Water Supply, Government of Nepal, Nepal.

Henning Bjornlund, UniSA Business, University of South Australia, Australia.

Louise Blessington, Australian National University, Australia.

Cristian Chadwick, Facultad de Ciencias Forestales y de la Conservación de la Naturaleza, Universidad de Chile, Chile.

Mario Chilundo, University Eduardo Mondlane, Mozambique.

Evan W. Christen, Penevy Services Pty Ltd, Australia.

Bethany Cooper, UniSA Business, University of South Australia, Australia.

Lin Crase, UniSA Business, University of South Australia, Australia.

Arnaud de Bonviller, ISL Ingénierie, France.

Simon de Bonviller, GESTE laboratory, ENGEES – University of Strasbourg, France.

Wilson de Sousa, National Institute for Irrigation, Mozambique.

Guillermo Donoso, Centro de Derecho y Gestión de Aguas, Pontifica Universidad Católica de Chile, and Departamento Economía Agraria, Facultad de Agronomía e Ingeniería Forestal, Pontificia Universidad Católica de Chile, Chile.

Dustin Garrick, University of Oxford, United Kingdom.

Gina Gilson, Smith School of Enterprise and the Environment, University of Oxford, United Kingdom.

R. Quentin Grafton, Crawford School of Public Policy, Australian National University, Australia and Lee Kuan Yew School of Public Policy, National University of Singapore, Singapore.

Juliane Haensch, Faculty of the Professions, University of Adelaide, Australia.
Qiuqiong Huang, Department of Agricultural Economics and Agribusiness, University of Arkansas, USA.
Andrew Johnson, Andrew Johnson & Associates, Adelaide, Australia.
Krasposy Kujinga, WaterNet, Zimbabwe.
Adam Loch, Faculty of the Professions, University of Adelaide, Australia.
Sophie Lountain, UniSA Business, University of South Australia, Australia.
Emmanuel Manzungu, University of Zimbabwe, Zimbabwe.
Matthew McCartney, International Water Management Institute, Colombo, Sri Lanka.
Makarius Mdemu, Ardhi University, Tanzania.
C. Dionisio Pérez-Blanco, Universidad de Salamanca, Spain, and Euro-Mediterranean Centre on Climate Change and Ca' Foscari University, Italy.
Jamie Pittock, Australian National University, Australia.
Kate Reardon-Smith, Centre for Applied Climate Sciences, University of Southern Queensland, Australia.
Lisa-Maria Rebelo, International Water Management Institute, Colombo, Sri Lanka.
Daniela Rivera, Centro de Derecho y Gestión de Aguas, Pontifica Universidad Católica de Chile, and Facultad de Derecho, Pontificia Universidad Católica de Chile, Chile.
Maheswor Shrestha, Water and Energy Commission Secretariat, Government of Nepal, Nepal.
Nancy E. Smith, Water Resource East, Enterprise Centre, University of East Anglia, United Kingdom.
Tianhe Sun, Collaborative Innovation Center for Beijing–Tianjin–Hebei Integrated Development, Hebei University of Economics and Business, China.
Julia Talbot-Jones, School of Government, Victoria University of Wellington, New Zealand.
André van Rooyen, International Crops Research Institute for the Semi-Arid Tropics, Zimbabwe.
Jinxia Wang, China Center for Agricultural Policy, School of Advanced Agricultural Sciences, Peking University, China.
Sarah Ann Wheeler, Faculty of the Professions, University of Adelaide, Australia.
Ying Xu, Faculty of the Professions, University of Adelaide, Australia.
Mike Young, Faculty of the Professions, University of Adelaide, Australia.
Alec Zuo, Faculty of the Professions, University of Adelaide, Australia.

Acknowledgements

First of all, I am very grateful to the many water policy experts from around the world who contributed their time and thoughts to writing about water scarcity, water policy and the presence of water markets in their regions for this book. I would like to acknowledge the help of a number of individuals in the creation of this book: Juliane Haensch and Adam Wheeler helped with formatting and editing; Adam Loch played an important role in the original idea for the book and also proof-read some chapters; and although numerous other colleagues have provided ideas and intellectual support, I would particularly like to thank Quentin Grafton and Dustin Garrick for their support and feedback. Funding for research in this book was provided through two Australian Research Council grants [DP200101191 and FT140100773].

Glossary

Adaptation	The response to major changes in the environment (e.g. global warming) and/or political and economic shocks. Adaptation is often imposed on individuals and societies by external undesirable changes.
Adoption (in agriculture)	A change in practice or technology.
Annual crops	Crops that go through their entire lifecycle in one growing season (e.g. cotton, rice, cereal).
Cap	A physical limit on the amount of water extractions that can be taken within a given water resource.
Carry-over	Arrangements which allow water licence holders to hold water in storages (water allocations not taken in a water accounting period) so that it is available in subsequent years.
Catchment (river valley)	An area determined by topographic features, within which rainfall contributes to run-off at a particular point.
Consumptive water use	The use of water for private benefit (e.g. irrigation, industry, urban, and stock and domestic uses).
Demand-side water management	The use of instruments such as regulation, education, property rights (water markets), prices and planning to influence water extraction or management.
Groundwater	The supply of freshwater found beneath the earth's surface (typically in aquifers).
Over-allocation	The total volume of water able to be extracted by the holders of water (access) entitlements at a given time exceeds the environmentally sustainable level of take for a water resource.
Permanent crops	Trees or shrubs, not grown in rotation, but occupying the soil and yielding harvests for several (usually more than five) consecutive years. Permanent crops mainly consist of fruit and berry trees, bushes, vines and olive trees and generally yield a higher added value per hectare than annual crops.
Regulated river system	Rivers regulated by major water infrastructure, such as dams, to supply water for varies uses.
Reliability	The frequency with which water allocated under a water (access) entitlement is able to be supplied in full.
Supply augmentation	Use of technology and infrastructure (eg dams, reservoirs, pumps) to address water scarcity.
Surface water	Water that flows over land and in watercourses or artificial channels.

Sustainable diversion limit	Maximum amount of water that can be taken for consumptive use reflecting an environmentally sustainable level of take (i.e. extractions must not compromise key environmental assets, ecosystem functions or productive base).
Transboundary water	A body of water that is shared by or forms the boundary between two or more political jurisdictions.
Unbundling	The legal separation of rights to land and rights to access water, have water delivered, use water on land or operate water infrastructure, all of which can be traded separately.
Unregulated river system	Rivers without major storages or rivers where the storages do not release water downstream.
Water allocation	A specific volume of water allocated to water (access) entitlements in a given season, according to the relevant water plan and the water availability in the water resource in that season (also known as temporary water).
Water licence (also known as water entitlement or water right)	A perpetual or ongoing licence to exclusive access to a share of water from a specified consumptive pool as defined in any relevant water plan (also known as permanent water).
Water markets (quantity)	Where there is significant exchange of the buying and selling of water rights. Can involve different types of rights: temporary; permanent; carrover; storage space; delivery, etc. A formal water market is one that has public property rights where water extraction rights are divisible, transferable, privately managed that can be bought or sold (in whole or part).
Water quality markets	Regulation and cap of point source emissions (e.g. nutrients) into watercourses and allows trade of credits between emitters.
Water trade	Where water is exchanged between a minimum of two parties, either formally or informally. Informal water trade can include arrangements between neighbours or swapping of water and other conditions.
Willingness to pay/ accept	The acceptable bid amount that an individual is prepared to pay/receive for acquiring/giving up the good in question.
WMRA (water market readiness framework)	A framework that outlines the three steps of functioning water markets, and provides a set various institutional factors that are needed for water markets to maximise community wellbeing.

1. Introduction to *Water Markets*: an overview and systematic literature review

Sarah Ann Wheeler and Ying Xu

1.1   PREFACE

This book, *Water Markets: A Global Assessment*, includes chapters written by water policy expert scholars highlighting the extent of water markets in 20 countries (and across 28 country areas/basins) around the world. Our case studies include countries within Africa, Asia, Europe, North America, Oceania and South America. The aims of this book are: (1) to provide information on the adoption of water markets around the world; (2) to clarify the range of contexts and issues in which water markets successfully emerge (through the application of a water market readiness framework); and (3) to provide practical insights and guidance on water markets for water managers, academics and research students, industry organisations and government policy-makers. This chapter provides an introduction and a systematic review of the water market literature to date.

1.2   BACKGROUND

There are grave concerns regarding future trends in water scarcity across the world, and the challenge of reconciling water supply and demand will intensify as water extractions increase (Barbier, 2019; Grafton et al., 2018). Water issues are often described as a 'wicked problem' because they have multiple, interconnected causes; whilst having many possible solution perspectives (Grafton and Wheeler, 2015; Quiggin, 2001). Hanemann (2006) describes nine economic aspects of water that presents issues for its management: (1) water has both private and public good aspects; (2) water's mobility; (3) water's variability; (4) cost of water; (5) price of water; (6) essentialness of water; (7) heterogeneity of water; (8) fallacy/misconception of using average value; and (9) benefits of water.

Policy-makers have a range of options to improve water management and allocation, and these very broadly fall into water supply augmentation and water demand management. Supply augmentation – otherwise known as 'hard' infrastructure or engineering solutions to increase water supply (for example, dam, irrigation infrastructure and weir construction) or substitution (for example, desalinated water) – has traditionally been the most promoted because it offers a technical and relatively rapid (and occasionally efficient where there are low marginal costs) method to address water scarcity (Wheeler and Garrick, 2020). Water demand-side management – otherwise known as 'soft' infrastructure and governance – includes educational measures (for example, information and campaigns), regulatory and/or planning processes (for example, legislation and regulation) and economic incentives (for example, economic pricing, subsidies and/or property right changes that allow water markets) (Wheeler and Garrick, 2020). Ideally both demand and supply responses should be integrated to address water security; however, this is frequently not the case (Barbier, 2019; Griffin, 2006; Sadoff et al., 2015).

Given that the choice of cost-effective supply augmentation projects is diminishing around the world, increasingly water demand management, and in particular water markets, will be further considered and implemented to address water scarcity and quality issues. However, little is known about the applicability of water markets, and what advice practitioners can use to assess how suitable their water resource situation is for water markets.

Water markets are often identified as an important way forward for addressing scarce water supply and increasing demand (Gómez Gómez et al., 2018). Previous books on water markets include (but are not limited to) Easter et al. (1998), Easter and Huang (2014) and Maestu (2013). Economists in particular believe in the power of markets to effectively and efficiently reallocate resources at the margin of use, and many argue that water trading is currently underutilised as a management option. On the other hand, there are many critics of water markets, arguing that 'water is too different to sell' (Griffin et al., 2013, p. 2). Griffin et al. (2013, p. 3) summarise this view of water trading opponents as espousing a water-is-different view that exaggerates biological requirements for water, confuses capital scarcity with water scarcity, seeks to preserve agrarian economies and claims public entitlement to water. On the other hand, Griffin et al. (2013) state that some economists over-emphasise the 'water is no different from other commodities' argument, and have too much faith in the market system resulting in an efficient economy.

We agree with Griffin et al. (2013) that water is sufficiently different to other commodities to mean that water markets are not automatically a panacea, and that certain water market failures may mean a net cost for society. There is a need for a variety of conditions to be met for water markets to be successful.

Indeed, water markets are not an easily adopted or implemented solution, often requiring significant time periods and political/financial investment.

1.3   WHAT IS WATER TRADING?

Water trading occurs in water markets, and can be defined as the process of buying and selling water licences (also called entitlements or rights). Three broad types of water trading can be defined: (1) short-term or temporary transfers of water (also known as water allocation trade); (2) medium-term leasing of water allocations to secure access to water for a period of time specified in a contract (also known as water leasing); and (3) permanent transfers of water entitlements – namely: (a) the ongoing property right to either a proportion or fixed quantity of the available water at a given source (also known as water entitlement trading); and (b) water delivery rights, that is, the right to have water delivered (Wheeler and Garrick, 2020).

In short, formal water markets involve reforming water law to transform water public property rights to a situation where some water use rights are divisible, transferable, privately managed and can be bought or sold (in whole or part) (Griffin et al., 2013). Formal water trading arrangements may comprise sanctioned rules, processes, catchment areas managed by governments and/or communities. Hence, water trading can change the location, timing and technical efficiency of water use (Easter et al., 1999; Howe et al., 1986). Formal water markets have evolved widely in the world since the 1970s (Chong and Sunding, 2006), and exist in differing stages across many countries in the world, as illustrated in this global water markets book. Formal water markets can be slow to develop in some regions for a number of reasons, such as local political circumstances and the interrelated nature of water use (for example, return flows) (Vaux and Howitt, 1984; Young, 1986).

Water markets can also be established informally, which typically involves water users located in a specific region or sharing a water resource. Informal water trading arrangements, for example, can include arrangements between neighbours and come in diverse forms (for example, private tube-wells in India or informal swapping of water in the United States, Spain and Mexico) (De Stefano and Hernández-Mora, 2016; Mukherji, 2008).

Another form of water markets includes water quality trading (which is similar to carbon emission trading). It involves regulation of the discharge of point sources (for example, industrial or urban) and allows trading of credits between polluters, with money generated often being used to encourage activities to reduce discharge of nutrients from non-point sources (for example, agriculture). The remainder of this book concentrates mainly on water quantity markets, but for further discussion on water quality markets see Uchida et al. (2018) and Leonard et al. (2019).

1.4   EVALUATING WATER MARKETS

As elaborated further in Chapter 2, there are many ways to evaluate the performance of water markets. Water allocation regimes, such as water markets, typically aim to comply with economic efficiency terms (focusing on wealth creation by a resource) and social equity considerations (focusing on the wealth distribution among sectors and individuals) (Dinar et al., 1997). The three distinct forms of economic efficiency associated with water markets are:

1. Allocative efficiency: improving water resource short-term decision-making to help reflect seasonal conditions (for example, weather, commodity price adjustments, cropping choices) is facilitated by water temporary trade.
2. Dynamic efficiency: improving water resource structural or long-term decision-making to reflect new investment opportunities, regulatory shifts in access arrangements (for example, extraction limits or embargos) or personal strategic choices (for example, retirement), which is best achieved through water permanent trade.
3. Productive efficiency: water price changes (both temporary and permanent) offer incentives for the efficient use of water resources as either an investment or input for productive outcomes.

Grafton et al. (2011) introduced an integrated framework to assess and compare the institutional foundations, economic efficiency and environmental sustainability of water markets around the world. The framework highlights important linkages between water market development, institutional constraints and management goals, and helps to identify which water markets contribute to integrated water resource management, which features of water markets require further development and how water governance can be improved (Grafton et al., 2011).

Grafton et al. (2016) provide extensive commentary about various arguments for and against water markets, including: privatisation (allocation of individual rights), deregulation (diminishment of the regulatory role of public organisations), decentralisation (transfer of decision-making and responsibility to a subsidiary level of authority), corporatisation (shift from public to corporate ownership), commercialisation (adoption of business models of practice and decision-making), marketisation (use of markets to determine use) and resource commodification (treatment of natural resources, including water, as a market good). Often those who argue against markets do so because they believe markets are a tool of global capitalism that results in appropriative privatisation, where state or private actors obtain water resources (without meaningful compensation) previously held in common ownership. However, as Griffin et al. (2013) and Grafton et al. (2016) emphasise, there is a clear need to distinguish between infrastructure privatisation and water licence privatisation. Also, while there may be examples of privatisation leading to appropriation, the issue that must be focused upon is whether this is the market's fault, or whether it is a result of the institutions that surround markets and market failure. Understanding market failure and preconditions needed for water markets is critical for their success.

1.5 WHAT HAS THE WATER MARKET LITERATURE STUDIED TO DATE?

To understand the focus and overview of the water market literature around the world, we conducted a search of the literature from January 1970 to December 2019 in the electronic databases Web of Science, ScienceDirect, ProQuest Central, Earth, Atmospheric and Aquatic Science Database and Google Scholar. The phrases 'water market', 'water trading', 'water trade', 'groundwater markets', 'water rights', 'water entitlements' and 'water allocation' were used as search terms for title, keywords and anywhere in the text. A flow chart of the literature selection process is included in Figure 1.1.

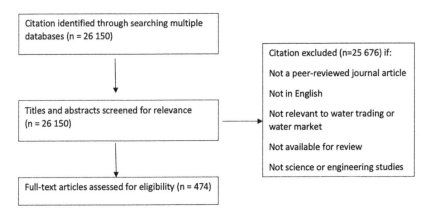

Figure 1.1   *Systematic review of water market literature, 1970–2019*

Our search was restricted to published English language studies, peer-reviewed journal articles relevant to water market/trading. After further excluding science and engineering studies and those focusing on saltwater or bottled water, there were 465 highly relevant studies on freshwater markets (including both surface water and groundwater) from social science perspectives. These 465 selected publications are further analysed below.

Figure 1.2 shows the trend of publication numbers of all water market/ trade-related studies and studies titled with 'water market/trade'. The number of all water market/trade-related studies increases significantly from 1970 to 2019, but the share of studies titled with 'water market/trade' increases only slightly every year. Since the mid-1980s, the share of studies titled with 'water market/trade' has grown less than proportionally to the total number of water market-related studies, suggesting a diversification of academic investigations from water markets per se to a broader variety of related issues.

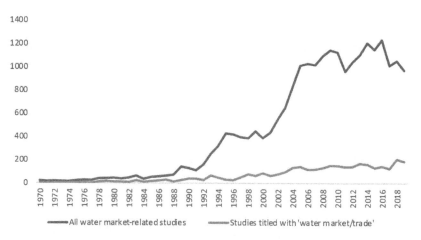

Figure 1.2   Cumulative number of studies in the water market/trade related literature, 1970–2019

Figure 1.3 shows the publication number trend. The annual average number of water market-related studies has increased from 0.4 in the 1970s to 23.2 in the 2010s. In the 2000s, the annual number of water market/trade-related studies increased significantly.

Most of these publications focus on water market issues in Australia (164 studies), which is the driest continent in the world. The United States (mostly the western states) receives the second-largest share of this literature (128 studies). China also has a significant share of the literature (33 studies), most of which were published in the 2010s. Studies on these three countries consist of more than 71 per cent of the selected publications.

Figure 1.4 presents the regional distribution of study contexts, where general water market investigations (without a specific country focus) are excluded. It is seen that most studies are on countries where certain regions face significant water resource constraints.

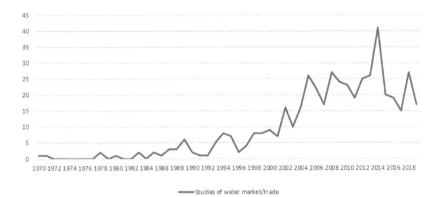

Figure 1.3   *Number of annual relevant studies in the water market/trade literature, 1970–2019*

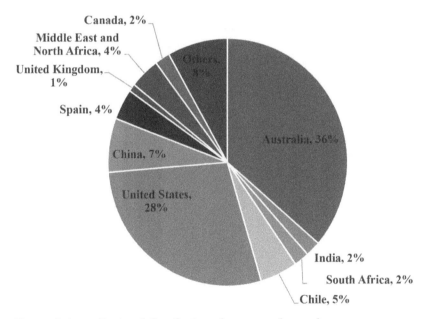

Figure 1.4   *Regional distribution of water market study contexts, 1970–2019*

The type of water market (for example, rural, urban, other – no specific focus) was identified from 458 publications: 259 studies investigated rural water markets, focusing particularly on agricultural water use and management; 22

studies studied water demand/consumption issues in urban areas; while 71 studies jointly considered both rural and urban areas to investigate water distribution issues among different production/consumption sectors. The remainder had no specific rural or urban focus.

Regarding analytical approaches, in our review of the water market literature, 210 studies (43.9 per cent) used qualitative methods and 261 studies (54.6 per cent) employed quantitative tools including theoretical economic modelling, optimization, simulation and econometric methods. A mixed-methods approach was used by 0.4 per cent of studies; and 1.1 per cent were literature syntheses.

Water market studies are published across a variety of disciplines. According to the Web of Science journal classifications, popular journal categories included Water Resources (253 studies), Economics (122 studies), Environmental Sciences (45 studies), Agricultural Economics and Policy (76 studies), Law (27 studies) and Development Studies (24 studies), among others (double-counting applies, as most journals are classified into two or more categories). Popular outlets included: *Water Policy* (27 studies), *Water Resources Research* (25 studies), *Agricultural Water Management* (23 studies), *Australian Journal of Agricultural and Resource Economics* (18 studies), *Journal of Hydrology* (17 studies), *Ecological Economics* (14 studies), *International Journal of Water Resources Development* (12 studies), *Natural Resources Journal* (11 studies), *American Journal of Agricultural Economics* (11 studies), *Water International* (10 studies) and *Journal of Environmental Management* (9 studies), among others.

Popular research topics of selected publications are presented in Table 1.1. There has been a wide variety of different research questions investigated in the literature, from understanding important preconditions for water markets, to analysing water market price and volume drivers, to understanding net overall benefits from water markets.

1.6 THIS BOOK

Given the increasing interest in water markets and the growing water market literature, Wheeler et al. (2017; which is republished with minor changes in Chapter 2) sought to develop an assessment framework for evaluating the practical benefits and institutional bases for the implementation and sustainability of water markets to address scarcity issues, called the water market readiness assessment (WMRA) framework. The WMRA consisted of three institutional factors as a prerequisite for establishing water markets:

1.   Step one: enabling institutions. Defining the total resource pool available for consumptive use and hydrological factors of use; and evaluating

Table 1.1   *Popular research topics in the water market literature,*
1970–2019

Research topic	Typical approach	Some examples (not exhaustive)
Feasibility of introducing water market(s) and framework assessment	Qualitative	Zarour and Isaac, 1993; Becker and Zeitouni, 1998; Bjornlund and McKay, 2000; Bate, 2002; Bjornlund, 2003; Vasquez, 2008; Grafton et al., 2011; Akram, 2013; Kirsch and Maxwell, 2015; Grafton et al., 2016; Prieto, 2016; Wheeler et al., 2017; Petterini, 2018
Water market performance	Qualitative and quantitative	McCarl et al., 1999; Neuman and Chapman, 1999; Mahan et al., 2002; Newlin et al., 2002; Zekri and Easter, 2005; Pujol et al., 2006; Bauer, 2010; Culp et al., 2014; Grafton and Horne, 2014; Wheeler, 2014; Wheeler et al., 2014; Bauer, 2015; Leonard et al., 2019
Privatisation and marketisation of the water sector	Qualitative	Glennon, 2004; Borzutzky and Madden, 2013; Glennon, 2015; Grafton et al., 2016
Water market policy evaluation	Qualitative and quantitative	Rosegrant et al., 1995; Brennan, 2006; van Heerden et al., 2008; Garrick and Aylward, 2012; Garrick et al., 2013; Jamshidi et al., 2016
Water demand and price analysis	Quantitative	Zarnikau, 1994; Saleth and Dinar, 2001; Ipe and Bhagwat, 2002; Yoskowitz, 2002; Garcia et al., 2005; Gulyani et al., 2005; Zilberman and Schoengold, 2005; Pullen and Colby, 2008; Wheeler et al., 2008a; Zuo et al., 2019: Schwabe et al., 2020
Farmers' willingness to pay for water or participate in water market	Quantitative	Saleth and Dinar, 2001; Ranjan and Shogren, 2006; Giannoccaro et al., 2015; Venkatachalam, 2015; Jaghdani and Brümmer, 2016; Wheeler et al., 2009, 2010
Human behaviour in water markets	Quantitative (experimental)	Lefebvre et al., 2012; Broadbent et al., 2014; Hansen et al., 2014; Nauges et al., 2016
Institutional arrangements and transaction costs	Qualitative and quantitative	Howitt, 1994; Shatanawi and Al-Jayousi, 1995; Nieuwoudt, 2000; Carey et al., 2002; Hadjigeorgalis and Lillywhite, 2004; Zhang et al., 2009; Zhao et al., 2013; Erfani et al., 2014; Breviglieri et al., 2018; Loch et al., 2018
Case studies: water market successes and failures	Qualitative and quantitative	Burness and Quirk, 1980; Yoskowitz, 1999; Bakker, 2002; Zegarra, 2002; Bauer, 2010; Zavalloni et al., 2014; Bauer, 2015
Water trade modelling	Quantitative	Louw and van Schalkwyk, 2001; Turral et al., 2005; Zaman et al., 2009; Alevy et al., 2010; Loch et al., 2011; Wittwer, 2011; Hung et al., 2014; Regnacq et al., 2016; Wheeler et al., 2008a, 2008b; Zuo et al., 2019

Research topic	Typical approach	Some examples (not exhaustive)
Design of water rights	Qualitative and quantitative	Johnson, 1971; Shupe et al., 1989; Michelsen and Young, 1993; Rosegrant and Binswanger, 1994; Ríos and Quiroz, 1995; Matthews, 2004; Solanes and Jouravlev, 2006; Whitford and Clark, 2007; Donohew, 2009; McKenzie, 2009; Nordblom et al., 2011; Raffensperger, 2011; Lefebvre et al., 2012; Jamshidi et al., 2016; Young, 2019
Inter-state water governance	Qualitative	Utton, 1985; Rodgers, 1986; Wheeler, 2014
Water use efficiency	Quantitative	Srivastava et al., 2009; Manjunatha et al., 2011; Razzaq et al., 2019
Environmental impacts	Quantitative	Tisdell, 2001; Lee et al., 2007; Rambonilaza and Neang, 2019
Climate impacts on water markets	Qualitative and quantitative	Pullen and Colby, 2006; Adler, 2008; Wheeler et al., 2008a, 2008b, 2013; Kahil et al., 2015; Ghosh, 2019; Zuo et al., 2019
Indigenous water rights/markets	Qualitative and quantitative	Nikolakis et al., 2013; von der Porten and de Loë, 2014; Nikolakis and Grafton, 2015; Taylor et al., 2016; Poirier and Schartmueller, 2012
Water quality trade	Qualitative and quantitative	Uchida et al., 2018; Leonard et al., 2019
Informal water markets	Qualitative and quantitative	Brozovic et al., 2002; Garrick et al., 2019; Munala and Kainz, 2012; Sima et al., 2013; Venkatachalam, 2015; Jaghdani and Brümmer, 2016; Razzaq et al., 2019

the current institutional, legislative, planning and regulatory capacity to facilitate water trade, involving: (a) specifying each resource share in perpetuity while allowing for changes in the proportion allocated to each share (comprises setting caps and regulations on use); (b) fully assigning responsibility for managing supply risk to users; (c) ensuring enforcement, strict regulation of caps and monitoring/compliance; and (d) keeping transaction costs low.

2. Step two: facilitating gains from trade. Developing clear and consistent trading rules; assessing benefits and costs of market-based reallocation; for example, numbers of individuals who can trade (versus adoption of trade); homogeneity of water use, adaptation benefits, cost of water reform, ongoing trade transaction costs and assessment of externalities. There is a difference between legislating for water trade to occur, allowing transfers between a small number of individuals, versus broader water reform legislation (for example, creating water registers with transparent,

complete and fully accessible data, clearer trade rules, public information sources).

3. Step three: monitoring and enforcement. Use of water markets and water extractions need ongoing monitoring and enforcement to ensure compliance, as well as continued development of trade-enabling mechanisms, including: seeking to limit/reduce transaction costs, scanning for unanticipated externalities, developing new market products (for example, option contracts or forwards) and then implementing, if needed, new legislative changes and planning requirements. Water market rules need flexibility to ensure water security and manage future uncertainty (Wheeler et al., 2017).

The institutional framework factors outlined in steps one and three above are needed for any property rights regime, while step two lists specific institutional factors required for water markets in particular. Given that water markets are complex economic instruments to design, develop, implement and sustain over time, and the fact that many countries have started to consider water markets as a management tool, there is a need for further testing of the WMRA framework to validate its robustness and practical usefulness in the field, particularly in developing-nation contexts with relatively weak institutional arrangements.

The purpose of this book is thus to apply the WMRA framework in a wider range of contextual applications across as many different countries as possible. This extended WMRA application is aimed at providing detailed insights and lessons for other countries going forward. The country case study chapters sought to apply the WMRA framework in different contexts and wherever possible to identify (as examples): (1) the local policy and practical management attitudes toward water markets, and whether there is any understanding/willingness to explore water market capacity to deal with river basins under scarcity stress; (2) relevant stages of water reforms toward market-based economic instruments, and whether other approaches might be more feasible/appropriate; and (3) detailed examinations of particular aspects of the WMRA framework to identify current gaps or requirements for reform if markets are to be established and sustained over time.

1.6.1 Water Market Case Studies Explored in this Book

The chapters written by water policy expert scholars in this book span six continents, across 20 countries. Chapter 2 presents the original journal article of the WMRA framework, and this journal article undertook case studies of Tasmania, Spain and the Diamond Valley in the United States of America (USA). Following Chapter 2, the chapters are arranged by continent, namely: Africa, Asia, Europe, South America, North America and Oceania.

More specifically, in Chapter 3 Jamie Pittock and co-authors explore the presence and use of water markets in Mozambique, Tanzania and Zimbabwe. Alec Zuo and co-authors apply the WRMA framework to water markets in Zhangye City, China in Chapter 4; while Sophie Lountain and co-authors look at groundwater markets in West Bengal, India in Chapter 5. In Chapter 6, Kate Reardon-Smith and co-authors investigate water governance conditions in the Lower Mekong Basin, in particular for the countries of Myanmar, Lao, Cambodia, Thailand and Vietnam. In Chapter 7, Andrew Johnson and co-authors discuss Nepal's water management challenges; and in Chapter 8 Irfan Baig and co-authors discuss groundwater markets in the Indus Basin Irrigation System, Pakistan.

Chapter 9 by Simon and Arnaud de Bonviller discusses the extent to which water markets are implemented in France in the areas of Poitou Marsh Basin and the Neste system; and in Chapter 10 Carlos Pérez-Blanco provides an overview of the development of water markets in Italy, particularly focussing on the Po River Basin District. Chapter 11 by Rosalind Bark and Nancy Smith looks at water markets in England. Chapter 12 by Guillermo Donoso and co-authors summarises Chile's experiences with water markets.

Chapter 13 by Gina Gilson and Dustin Garrick looks at the growth of environmental water markets in the Columbia Basin (USA); while in Chapter 14 Julia Talbot-Jones and Quentin Grafton study water markets in Canterbury, New Zealand.

Chapter 15 summarises the key lessons learnt from all the case studies that have applied the WMRA framework, as well as providing a quantitative overview of the progress towards fully functioning water markets around the world.

Overall, we hope that this book will allow water professionals to more fully comprehend the practical issues associated with assessing water market implementation and, if desired and feasible, how to ensure their continued development for improved societal well-being and water allocation over time.

REFERENCES

Adler, J.H. (2008). Water marketing as an adaptive response to the threat of climate change. *Hamline Law Review, 31*(3), 729–754.

Akram, A.A. (2013). Is a surface-water market physically feasible in Pakistan's Indus Basin Irrigation System? *Water International, 38*(5), 552–570.

Alevy, J.E., Cristi, O., and Melo, O. (2010). Right-to-choose auctions: a field study of water markets in the Limari valley of Chile. *Agricultural and Resource Economics Review, 39*(1203-2016-95457), 213–226.

Bakker, K. (2002). From state to market? Water mercantilización in Spain. *Environment and Planning A, 34*(5), 767–790.

Barbier, E. (2019). *The Water Paradox: Overcoming the Global Crisis in Water Management*. Yale University Press.

Bate, R. (2002). Water: can property rights and markets replace conflict. In J. Morris (ed.), *Sustainable Development: Promoting Progress or Perpetuating Poverty* (pp. 1–16). Profile Books.

Bauer, C. (2010). Market approaches to water allocation: lessons from Latin America. *Journal of Contemporary Water Research and Education, 144*(1), 44–49.

Bauer, C.J. (2015). Water conflicts and entrenched governance problems in Chile's market model. *Water Alternatives, 8*(2), 147–172.

Becker, N., and Zeitouni, N. (1998). A market solution for the Israeli–Palestinian water dispute. *Water International, 23*(4), 1–5.

Bjornlund, H. (2003). Efficient water market mechanisms to cope with water scarcity. *Water Resources Development, 19*(4), 553–567.

Bjornlund, H., and McKay, J. (2000). Do water markets promote a socially equitable reallocation of water? A case study of a rural water market in Victoria, Australia. *Rivers, 7*(2), 141–154.

Borzutzky, S., and Madden, E.F. (2013). Markets awash: the privatization of Chilean water markets. *Journal of International Development, 25*(2), 251–275.

Brennan, D. (2006). Water policy reform in Australia: lessons from the Victorian seasonal water market. *Australian Journal of Agricultural and Resource Economics, 50*(3), 403–423.

Breviglieri, G.V., do Sol Osório, G.I., and Puppim de Oliveira, J.A. (2018). Understanding the emergence of water market institutions: learning from functioning water markets in three countries. *Water Policy, 20*(6), 1075–1091.

Broadbent, C.D., Brookshire, D.S., Coursey, D., and Tidwell, V. (2014). An experimental analysis of water leasing markets focusing on the agricultural sector. *Agricultural Water Management, 142*, 88–98.

Brozovic, N., Carey, J.M., and Sunding, D.L. (2002). Trading activity in an informal agricultural water market: an example from California. *Water Resources Update, 121*(1), 3–16.

Burness, H.S., and Quirk, J.P. (1980). Water law, water transfers, and economic efficiency: the Colorado River. *Journal of Law and Economics, 23*(1), 111–134.

Carey, J., Sunding, D.L., and Zilberman, D. (2002). Transaction costs and trading behavior in an immature water market. *Environment and Development Economics, 7*(4), 733–750.

Chong, H., and Sunding, D. (2006). Water markets and trading. *Annual Review of Environmental Resource Economics, 31*, 239–264.

Culp, P.W., Glennon, R.J., and Libecap, G. (2014). *Shopping for Water: How the Market can Mitigate Water Shortages in the American West*. Island Press.

De Stefano L., and Hernández-Mora N. (2016). Los mercados informales de aguas en españa: una visión de conjunto. In J. Gómez-Limón and J. Calatrava (eds), *Los Mercados de Agua en España: Presente Y Perspectivas* (pp. 95–121). Cajamar Caja Rural, Amería.

Dinar, A., Rosegrant, M.W., and Meinzen-Dick, R. (1997). Water allocation mechanisms: principles and examples. Policy Research Working Paper, World Bank, no. 1779. http://elibrary.worldbank.org/doi/abs/10.1596/1813-9450-1779.

Donohew, Z. (2009). Property rights and western United States water markets. *Australian Journal of Agricultural and Resource Economics, 53*(1), 85–103.

Easter, K., and Huang, Q. (eds) (2014). *Water Markets for the 21st Century*. Springer.

Easter, K., Rosegrant, M., and Dinar, A. (1998). *Markets for Water: Potential and Performance.* Natural Resource Management and Policy series. Kluwer Academic Publishers.

Easter, K., Rosegrant, M., and Dinar, A. (1999). Formal and informal markets for water: institutions, performance, and constraints. *World Bank Research Observer, 14*(1), 99–116.

Erfani, T., Binions, O., and Harou, J.J. (2014). Simulating water markets with transaction costs. *Water Resources Research, 50*(6), 4726–4745.

Garcia, S., Guérin-Schneider, L., and Fauquert, G. (2005). Analysis of water price determinants in France: cost recovery, competition for the market and operator's strategy. *Water Science and Technology: Water Supply, 5*(6), 173–181.

Garrick, D., and Aylward, B. (2012). Transaction costs and institutional performance in market-based environmental water allocation. *Land Economics, 88*(3), 536–560.

Garrick, D., De Stefano, L., Yu, W., Jorgensen, I., O'Donnell, E., et al. (2019). Rural water for thirsty cities: a systematic review of water reallocation from rural to urban regions. *Environmental Research Letters,* 14(4), 043003.

Garrick, D., Whitten, S.M., and Coggan, A. (2013). Understanding the evolution and performance of water markets and allocation policy: a transaction costs analysis framework. *Ecological Economics, 88,* 195–205.

Ghosh, S. (2019). Droughts and water trading in the western United States: recent economic evidence. *International Journal of Water Resources Development, 35*(1), 145–159.

Giannoccaro, G., Castillo, M., and Berbel, J. (2015). An assessment of farmers' willingness to participate in water trading in southern Spain. *Water Policy, 17*(3), 520–537.

Glennon, R. (2004). Water scarcity, marketing, and privatization. *Texas Law Review, 83,* 1873.

Glennon, R. (2015). Should farm communities support water markets? *Western Farm Press,* 8 May.

Gómez Gómez, C.M., Pérez-Blanco, C.D., Adamson, D., and Loch, A. (2018). Managing water scarcity at a river basin scale with economic instruments. *Water Economics and Policy, 4*(1), 1750004. DOI: 10.1142/S2382624X17500047.

Grafton, R.Q., and Horne, J. (2014). Water markets in the Murray–Darling basin. *Agricultural Water Management, 145*(C), 61–71.

Grafton, R.Q., and Wheeler, S.A. (2015). Water economics. In R. Halvorsen and D. Layton (eds), *Handbook on the Economics of Natural Resources* (pp. 401–420). Edward Elgar Publishing.

Grafton, R.Q., Horne, J., and Wheeler, S.A. (2016). On the marketisation of water: evidence from the Murray–Darling Basin, Australia. *Water Resources Management, 30*(3), 913–926.

Grafton, R.Q., Libecap, G., McGlennon, S., Landry, C., and O'Brien, B. (2011). An integrated assessment of water markets: a cross-country comparison. *Review of Environmental Economics and Policy, 5*(2), 219–239.

Grafton, R.Q., Williams, J., Perry, C.J., Molle, F., Ringler, C., et al. (2018). The paradox of irrigation efficiency. *Science, 361*(6404), 748–750.

Griffin, R.C. (2006). *Water Resource Economics: The Analysis of Scarcity, Policies, and Projects.* MIT Press.

Griffin, R.C., Peck, D.E., and Maestu, J. (2013). Introduction: myths, principles and issues in water trading. In J. Maestu (ed.), *Water Trading and Global Water Scarcity: International Experiences* (pp. 1–14). RFF Press Water Policy Series.

Gulyani, S., Talukdar, D., and Mukami Kariuki, R. (2005). Universal (non) service? Water markets, household demand and the poor in urban Kenya. *Urban Studies, 42*(8), 1247–1274.

Hadjigeorgalis, E., and Lillywhite, J. (2004). The impact of institutional constraints on the Limarí River Valley water market. *Water Resources Research, 40*(5), 752–1688.

Hanemann, W.H. (2006). The economic conception of water. In P.P. Rogers, M.R. Llamas and L. Martinez-Cortina (eds), *Water Crisis: Myth or Reality?* (pp. 61–91). Taylor & Francis.

Hansen, K., Kaplan, J., and Kroll, S. (2014). Valuing options in water markets: a laboratory investigation. *Environmental and Resource Economics, 57*(1), 59–80.

Howe, C., Schurmeier, D., and Shaw Jr, W. (1986). Innovative approaches to water allocation: the potential for water markets. *Water Resources Research, 22*(4), 439–445.

Howitt, R.E. (1994). Empirical analysis of water market institutions: the 1991 California water market. *Resource and Energy Economics, 16*(4), 357–371.

Hung, M.F., Shaw, D., and Chie, B.T. (2014). Water trading: locational water rights, economic efficiency, and third-party effect. *Water, 6*(3), 723–744.

Ipe, V.C., and Bhagwat, S.B. (2002). Chicago's water market: dynamics of demand, prices and scarcity rents. *Applied Economics, 34*(17), 2157–2163.

Jaghdani, T., and Brümmer, B. (2016). Determinants of willingness to pay for groundwater: insights from informal water markets in Rafsanjan, Iran. *International Journal of Water Resources Development, 32*(6), 944–960.

Jamshidi, S., Ardestani, M., and Hossein Niksokhan, M. (2016). A seasonal waste load allocation policy in an integrated discharge permit and reclaimed water market. *Water Policy, 18*(1), 235–250.

Johnson, D.D. (1971). An optimal state water law: fixed water rights and flexible market prices. *Virginia Law Review, 57*(3), 345–374.

Kahil, M.T., Dinar, A., and Albiac, J. (2015). Modeling water scarcity and droughts for policy adaptation to climate change in arid and semiarid regions. *Journal of Hydrology, 522*, 95–109.

Kirsch, B.R., and Maxwell, R.M. (2015). The use of a water market to minimize drought-induced losses in the Bay Area of California. *Journal – American Water Works Association, 107*(5), E274–E281.

Lee, L.Y., Ancev, T., and Vervoort, W. (2007). Environmental and economic impacts of water scarcity and market reform on the Mooki catchment. *Environmentalist, 27*(1), 39–49.

Lefebvre, M., Gangadharan, L., and Thoyer, S. (2012). Do security-differentiated water rights improve the performance of water markets? *American Journal of Agricultural Economics, 94*(5), 1113–1135.

Leonard, B., Costello, C., and Libecap, G.D. (2019). Expanding water markets in the western United States: barriers and lessons from other natural resource markets. *Review of Environmental Economics and Policy, 13*(1), 43–61.

Loch, A., Bjornlund, H., and McIver, R. (2011). Achieving targeted environmental flows: alternative allocation and trading models under scarce supply – lessons from the Australian reform process. *Environment and Planning C: Government and Policy, 29*(4), 745–760.

Loch, A., Wheeler, S.A., and Settre, C. (2018). Private transaction costs of water trade in the Murray–Darling Basin. *Ecological Economics, 146*, 560–573.

Louw, D.B., and Van Schalkwyk, H.D. (2001). The impact of transaction costs on water trade in a water market allocation regime. *Agrekon, 40*(4), 780–793.

Maestu, J. (ed.) (2013). *Water Trading and Global Water Scarcity: International Experiences*. RFF Press Water Policy Series.

Mahan, R.C., Horbulyk, T.M., and Rowse, J.G. (2002). Market mechanisms and the efficient allocation of surface water resources in southern Alberta. *Socio-Economic Planning Sciences*, *36*(1), 25–49.

Manjunatha, A.V., Speelman, S., Chandrakanth, M.G., and Van Huylenbroeck, G. (2011). Impact of groundwater markets in India on water use efficiency: a data envelopment analysis approach. *Journal of Environmental Management*, *92*(11), 2924–2929.

Matthews, O.P. (2004). Fundamental questions about water rights and market reallocation. *Water Resources Research*, *40*(9). DOI: 10.1029/2003WR002836.

McCarl, B.A., Dillon, C.R., Keplinger, K.O., and Williams, R.L. (1999). Limiting pumping from the Edwards Aquifer: an economic investigation of proposals, water markets, and spring flow guarantees. *Water Resources Research*, *35*(4), 1257–1268.

McKenzie, M. (2009). Water rights in NSW: properly property. *Sydney Law Review*, *31*, 443.

Michelsen, A.M., and Young, R.A. (1993). Optioning agricultural water rights for urban water supplies during drought. *American Journal of Agricultural Economics*, *75*(4), 1010–1020.

Mukherji, A. (2008). Spatio-temporal analysis of markets for groundwater irrigation services in India: 1976–1977 to 1997–1998. *Hydrogeology Journal*, *16*(6), 1077–1087.

Munala, G., and Kainz, H. (2012). Managing interactions in the informal water market: the case of Kisumu, Kenya. *Development in Practice*, *22*(3), 347–360.

Nauges, C., Wheeler, S.A., and Zuo, A. (2016). Elicitation of irrigators' risk preferences from observed behaviour. *Australian Journal of Agricultural Resource Economics*, *60*(3), 442–458.

Neuman, J.C., and Chapman, C. (1999). Wading into the water market: the first five years of the Oregon Water Trust. *Journal of Environmental Law and Litigation*, *14*(1), 135–184.

Newlin, B.D., Jenkins, M.W., Lund, J.R., and Howitt, R.E. (2002). Southern California water markets: potential and limitations. *Journal of Water Resources Planning and Management*, *128*(1), 21–32.

Nieuwoudt, W.L. (2000). Water market institutions: lessons from Colorado. *Agrekon*, *39*(1), 58–67.

Nikolakis, W., and Grafton, R.Q. (2015). Putting Indigenous water rights to work: the Sustainable Livelihoods Framework as a lens for remote development. *Community Development*, *46*(2), 149–163.

Nikolakis, W.D., Grafton, R.Q., and To, H. (2013). Indigenous values and water markets: survey insights from northern Australia. *Journal of Hydrology*, *500*, 12–20.

Nordblom, T.L., Reeson, A.F., Finlayson, J.D., Hume, I.H., Whitten, S.M., and Kelly, J.A. (2011). Price discovery and distribution of water rights linking upstream tree plantations to downstream water markets: experimental results. *Water Policy*, *13*(6), 810–827.

Petterini, F.C. (2018). The possibility of a water market in Brazil. *Economia*, *19*(2), 187–200.

Poirier, R., and Schartmueller, D. (2012). Indigenous water rights in Australia. *Social Science Journal*, *49*(3), 317–324.

Prieto, M. (2016). Bringing water markets down to Chile's Atacama Desert. *Water International*, *41*(2), 191–212.

Pujol, J., Raggi, M., and Viaggi, D. (2006). The potential impact of markets for irrigation water in Italy and Spain: a comparison of two study areas. *Australian Journal of Agricultural and Resource Economics*, *50*(3), 361–380.

Pullen, J.L., and Colby, B.G. (2008). Influence of climate variability on the market price of water in the Gila-San Francisco Basin. *Journal of Agricultural and Resource Economics*, *33*(3), 473–487.

Quiggin, J. (2001). Environmental economics and the Murray–Darling river system. *Australian Journal of Agricultural and Resource Economics*, *45*(1), 67–94.

Raffensperger, J.F. (2011). Matching users' rights to available groundwater. *Ecological Economics*, *70*(6), 1041–1050.

Rambonilaza, T., and Neang, M. (2019). Exploring the potential of local market in remunerating water ecosystem services in Cambodia: an application for endogenous attribute non-attendance modelling. *Water Resources and Economics*, *25*, 14–26.

Ranjan, R., and Shogren, J.F. (2006). How probability weighting affects participation in water markets. *Water Resources Research*, *42*(8), 1–10.

Razzaq, A., Qing, P., Abid, M., Anwar, M., and Javed, I. (2019). Can the informal groundwater markets improve water use efficiency and equity? Evidence from a semi-arid region of Pakistan. *Science of The Total Environment*, *666*, 849–857.

Regnacq, C., Dinar, A., and Hanak, E. (2016). The gravity of water: water trade frictions in California. *American Journal of Agricultural Economics*, *98*(5), 1273–1294.

Ríos, M.A., and Quiroz, J.A. (1995). The market of water rights in Chile: major issues. *Cuadernos de economía*, *32*(97), 317–345.

Rodgers, A.B. (1986). The limits of state activity in the interstate water market. *Land and Water Law Review*, *21*, 357.

Rosegrant, M.W., and Binswanger, H.P. (1994). Markets in tradable water rights: potential for efficiency gains in developing country water resource allocation. *World Development*, *22*(11), 1613–1625.

Rosegrant, M.W., Schleyer, R.G., and Yadav, S.N. (1995). Water policy for efficient agricultural diversification: market-based approaches. *Food Policy*, *20*(3), 203–223.

Sadoff, C., Hall, J.W., Grey, D., Aerts, J.C.J.H., Ait-Kadi, M., et al. (2015). *Securing Water, Sustaining Growth. Report of the GWP/OECD Task Force on Water Security and Sustainable Growth*. Oxford University.

Saleth, R.M., and Dinar, A. (2001). Preconditions for market solution to urban water scarcity: empirical results from Hyderabad City, India. *Water Resources Research*, *37*(1), 119–131.

Schwabe, K., Nemati, M., Landry, C., and Zimmerman, G. (2020). Water markets in the western United States: trends and opportunities. *Water*, *12*(233), 1–15.

Shatanawi, M.R., and Al-Jayousi, O. (1995). Evaluating market-oriented water policies in Jordan: a comparative study. *Water International*, *20*(2), 88–97.

Shupe, S.J., Weatherford, G.D., and Checchio, E. (1989). Western water rights: the era of reallocation. *Natural Resources Journal*, *29*, 413.

Sima, L.C., Kelner-Levine, E., Eckelman, M.J., McCarty, K.M., and Elimelech, M. (2013). Water flows, energy demand, and market analysis of the informal water sector in Kisumu, Kenya. *Ecological Economics*, *87*, 137–144.

Solanes, M., and Jouravlev, A. (2006). Water rights and water markets: lessons from technical advisory assistance in Latin America. *Irrigation and Drainage: The Journal of the International Commission on Irrigation and Drainage*, *55*(3), 337–342.

Srivastava, S.K., Kumar, R., and Singh, R.P. (2009). Extent of groundwater extraction and irrigation efficiency on farms under different water-market regimes in central

Uttar Pradesh. *Agricultural Economics Research Review*, 22(1). DOI: 10.22004/ ag.econ.57384.

Taylor, K.S., Moggridge, B.J., and Poelina, A. (2016). Australian Indigenous water policy and the impacts of the ever-changing political cycle. *Australasian Journal of Water Resources*, 20(2), 132–147.

Tisdell, J.G. (2001). The environmental impact of water markets: an Australian case-study. *Journal of Environmental Management*, 62(1), 113–120.

Turral, H.N., Etchells, T., Malano, H.M.M., Wijedasa, H.A., Taylor, P., et al. (2005). Water trading at the margin: the evolution of water markets in the Murray–Darling Basin. *Water Resources Research*, 41(7). DOI: 10.1029/2004WR003463.

Uchida, E., Swallow, S.K., Gold, A.J., Opaluch, J., Kafle, A., et al. (2018). Integrating watershed hydrology and economics to establish a local market for water quality improvement: a field experiment. *Ecological Economics*, 146, 17–25.

Utton, A.E. (1985). In search of an integrating principle for interstate water law: regulation versus the market place. *Natural Resources Journal*, 25(4), 985–1004.

van Heerden, J.H., Blignaut, J., and Horridge, M. (2008). Integrated water and economic modelling of the impacts of water market instruments on the South African economy. *Ecological Economics*, 66(1), 105–116.

Vasquez, J. (2008). Feasibility of a water market in Colombia. *International Journal of Sustainable Development and Planning*, 3(4), 394–400.

Vaux Jr, H.J., and Howitt, R.E. (1984). Managing water scarcity: an evaluation of interregional transfers. *Water Resources Research*, 20(7), 785–792.

Venkatachalam, L. (2015). Informal water markets and willingness to pay for water: a case study of the urban poor in Chennai City, India. *International Journal of Water Resources Development*, 31(1), 134–145.

von der Porten, S., and de Loë, R.C. (2014). Water policy reform and Indigenous governance. *Water Policy*, 16(2), 222–243.

Wheeler, S.A. (2014). Insights, lessons and benefits from improved regional water security and integration in Australia. *Water Resources and Economics*, 8, 57–78.

Wheeler, S.A., and Garrick, D. (2020). A tale of two water markets in Australia: lessons for understanding participation in formal water markets. *Oxford Review of Economic Policy*, 36(1), 132–153.

Wheeler, S.A., Bjornlund, H., Shanahan, M., and Zuo, A. (2008a). Factors influencing water allocation and entitlement prices in the Greater Goulburn Area of Australia. In Y. Villacampa, C.A. Brebbia and D. Prats Rico (eds), *Sustainable Irrigation Management, Technologies and Policies II* (pp. 63–71). WIT Press.

Wheeler, S., Bjornlund, H., Shanahan, M., and Zuo, A. (2008b). Price elasticity of water allocations demand in the Goulburn-Murray Irrigation District. *Australian Journal of Agricultural and Resource Economics*, 52(1), 37–55.

Wheeler, S., Bjornlund, H., Shanahan, M., and Zuo, A. (2009). Who trades water allocations? Evidence of the characteristics of early adopters in the Goulburn–Murray Irrigation District, Australia 1998–1999. *Agricultural Economics*, 40(6), 631–643.

Wheeler, S., Bjornlund, H., Zuo, A., and Shanahan, M. (2010). The changing profile of water traders in the Goulburn–Murray Irrigation District, Australia. *Agricultural Water Management*, 97(9), 1333–1343.

Wheeler, S.A., Garrick, D., Loch, A., and Bjornlund, H. (2013). Evaluating water market products to acquire water for the environment in Australia. *Land Use Policy*, 30(1), 427–436.

Wheeler, S.A., Loch, A., Crase, L., Young, M., and Grafton, R.Q. (2017). Developing a water market readiness assessment framework. *Journal of Hydrology*, *552*, 807–820.

Wheeler, S.A., Loch, A., Zuo, A., and Bjornlund, H. (2014). Reviewing the adoption and impact of water markets in the Murray–Darling Basin, Australia. *Journal of Hydrology*, *518*, 28–41.

Whitford, A.B., and Clark, B.Y. (2007). Designing property rights for water: mediating market, government, and corporation failures. *Policy Sciences*, *40*(4), 335–351.

Wittwer, G. (2011). Confusing policy and catastrophe: buybacks and drought in the Murray–Darling Basin. *Economic Record*, *30*(3), 289–430.

Yoskowitz, D.W. (1999). Spot market for water along the Texas Rio Grande: opportunities for water management. *Natural Resources Journal*, *39*(2), 345–355.

Yoskowitz, D.W. (2002). Price dispersion and price discrimination: empirical evidence from a spot market for water. *Review of Industrial Organization*, *20*(3), 283–289.

Young, R.A. (1986). Why are there so few transactions among water users? *American Journal of Agricultural Economics*, *68*(5), 1143–1151.

Young, M. (2019). *Sharing Water: The Role of Robust Water-Sharing Arrangements in Integrated Water Resources Management*. Perspectives paper by Global Water Partnership. https://www.gwp.org/globalassets/global/toolbox/publications/perspective-papers/gwp-sharing-water.pdfhttps://www.gwp.org/globalassets/global/toolbox/publications/perspective-papers/gwp-sharing-water.pdf.

Zaman, A.M., Malano, H.M.,and Davidson, B. (2009). An integrated water trading-allocation model, applied to a water market in Australia. *Agricultural Water Management*, *96*(1), 149–159.

Zarnikau, J. (1994). Spot market pricing of water resources and efficient means of rationing water during scarcity (water pricing). *Resource and Energy Economics*, *16*(3), 189–210.

Zarour, H., and Isaac, J. (1993). Nature's apportionment and the open market: a promising solution to the Arab–Israeli water conflict. *Water International*, *18*(1), 40–53.

Zavalloni, M., Raggi, M., and Viaggi, D. (2014). Water harvesting reservoirs with internal water reallocation: a case study in Emilia Romagna, Italy. *Journal of Water Supply: Research and Technology – AQUA*, *63*(6), 489–496.

Zegarra, E. (2002). *Water Market and Coordination Failures: The Case of the Limari Valley in Chile*. University of Wisconsin.

Zekri, S., and Easter, W. (2005). Estimating the potential gains from water markets: a case study from Tunisia. *Agricultural Water Management*, *72*(3), 161–175.

Zhang, J., Zhang, F., Zhang, L., and Wang, W. (2009). Transaction costs in water markets in the Heihe River Basin in Northwest China. *International Journal of Water Resources Development*, *25*(1), 95–105.

Zhao, J., Cai, X., and Wang, Z. (2013). Comparing administered and market-based water allocation systems through a consistent agent-based modeling framework. *Journal of Environmental Management*, *123*, 120–130.

Zilberman, D., and Schoengold, K. (2005). The use of pricing and markets for water allocation. *Canadian Water Resources Journal*, *30*(1), 47–54.

Zuo, A., Qiu, F., and Wheeler, S. (2019). Examining volatility dynamics, spillovers and government water recovery in Murray–Darling Basin water markets. *Resource and Energy Economics*, *58*, 101113--1-101113-16.

2. Developing a water market readiness assessment framework[1]

Sarah Ann Wheeler, Adam Loch, Lin Crase, Mike Young and R. Quentin Grafton

2.1 INTRODUCTION

The supply of fresh water is finite and, in many locations, sufficiently scarce such that it is not possible to satisfy all competing uses. The challenge of reconciling supply and demand will intensify as global water extractions are expected to increase 55 per cent by 2050 (WWAP, 2014). This raises concerns about future trends in global water security; defined as the ability to safeguard access to water for livelihoods and development, to protect against water pollution and water-related disasters, and to preserve ecosystems to help ensure peace and political stability (UN Water, 2013). At the very least it requires approaches to water governance that explicitly acknowledge the need to manage risks and to live within biophysical limits, even if the precise management approaches are contested (Garrick and Hall, 2014).

There are two diverse arrangements for dealing with water scarcity risk and reallocation: demand-side management and supply augmentation. Demand-side management includes educational measures (for example, providing information on how to decrease water use in homes/farms), regulatory and/or planning processes (for example, legislative change coupled with catchment water-sharing plans or restrictions) and economic incentives (for example, pricing to discourage over-use, the use of subsidies that increase technical water use efficiency, and/or arrangements that allow the trading of limited opportunities to use water). Supply augmentation (for example, further dam and weir construction) or substitution (for example, desalinated water) has traditionally been used and promoted by managers because it offers a technical and relatively rapid 'fix' to address demand gaps. Ideally, both demand and supply responses should be integrated, but this is frequently not the case, highlighting the need for governance arrangements that better coordinate water demand and supply (Sadoff et al., 2015).

Three main water reallocation approaches are usually discussed in the literature: administrative, collective negotiations or agreements and market-based transfers (Meinzen-Dick and Ringler, 2008). Market-based reallocation may be unsuitable for developing economies due to less-clearly specified, distributed and prioritised water rights; uncertain operational rules; inadequate or disconnected supply and distribution infrastructure and poor water source, supply, usage and measurement data; and unequal access to the rule of law. In such places, less complex and costly non-market reallocation approaches may be more suitable (Marston and Cai, 2016); especially where a priority is placed on equity and delivery of water to the poor and vulnerable rather than efficiency considerations. In many developed economies growing water scarcity, greater environmental concern and limited supply-side options have driven an increased emphasis of demand-side reallocation policies, especially water pricing, charges and markets. Often, the establishment of the conditions that enable efficient trading and the eventual full emergence of markets is more accidental than planned (Griffin, 2006). Proponents of water markets argue that they offer more efficient and effective approaches to reallocate water, and can also protect social and environmental values (e.g., Chong and Sunding, 2006; Crase and O'Keefe, 2009). By contrast, others suggest that markets commodify water to benefit the wealthy and powerful at the expense of the most vulnerable, their communities or the environment (Barlow and Clarke, 2002). Market power, especially in economies where there are gross income inequalities, can drive negative perceptions of markets as a well-accepted means of reallocating water (Easter and Huang, 2014) and also increase the transaction costs to achieve market arrangements. Debate about water markets has also occasionally focused on the privatisation of urban water supplies (Goldman, 2007; Segerfeldt, 2005). Regardless of the viewpoint, achieving water security remains a major global challenge, and for this reason water markets still remain a possible policy response.

To address the water scarcity challenge, useful general frameworks exist for comparing water institutions (e.g., Dinar and Saleth, 2005) and also for proposing improvements to water market implementation and performance. These relate to issues of effective legal property rights (Tan, 2005), exchange frameworks for efficient transfers (Griffin, 2006), initial implementation recommendations (Maestu and Gómez Ramos, 2013) and agendas for comparison and performance evaluations across different water market contexts (Grafton et al., 2012). Nevertheless, despite significant attention devoted to the study of water rights/markets there is, until now, very little practical guidance to evaluate the appropriateness of water markets in emergent or semi-developed situations (Young, 2014b). Given the multi-decade experience of water markets in a number of countries, it is timely to distil insights about markets as a response to water scarcity. For the first time in the literature, we provide a framework that identifies the conditions necessary to facilitate reallocation via water trading; where that arrangement is perceived as an appropriate strategy. Our intent is for the framework to be used by water managers and planners to identify possible barriers to the implementation of water markets. In so doing, we stress that we do not propose water markets as a universal solution for the multidimensional problems of water security.

2.2   CONCEPTUAL BACKGROUND AND WATER MARKET DEVELOPMENT FACTORS

2.2.1   Water Trading

Water trading can be defined as the voluntary buying and selling of water in some quantifiable form; in either the present or future. In essence, there are three types of water trading: (1) short-term or temporary transfers of water that is already allocated and available for immediate use; (2) medium-term leasing of water allocations in a manner that enables a water user to plan secure access to water for a period of time; and (3) permanent transfers of water entitlements – the ongoing property right to either a proportion or a fixed quantity of the available water at a given source. Water trading arrangements can range from informal arrangements between neighbours (Maestu, 2013; Shah and Ballabh, 1997), to formal recognition and management by governments and/or communities. Formal government and/or community sanctioned water trading arrangements involve a variety of rules and processes designed to protect the interests of all users, including third parties who might otherwise be adversely effected (NWC, 2011a, 2011b). The formal trade of water thus has a number of possible benefits; including that it can help ensure that water use costs (and its opportunity costs), are explicitly accounted for by water users.

A key challenge for those interested in the role of water trading in helping to manage water use is the fact that water is an 'un-cooperative' commodity (Bakker, 2005, 2007). Its value is derived not only from its quantity, but also from its quality, reliability, timing, location and use. In many cases, trade involves reallocation: sometimes only to a neighbour as mentioned above, but increasingly to other sectors and even other regions (Grafton et al., 2010). Trade can – and often does – change who, where and how water is used, which can affect subsequent extraction by downstream parties or future use of aquifers. Thus, changes in the location, timing and technical efficiency of water use matter (Bauer, 2004; Easter et al., 1999; Howe et al., 1986; Young and McColl, 2009), as do the costs of trading and enforcing water rights (Garrick et al., 2013; McCann and Easter, 2004) and the effects on non-consumptive uses such as transport, hydro-power generation and the environment.

Considerable progress has been made in the development of water trading and marketing arrangements. Countries such as Australia, the United States, Chile, Mexico, South Africa and China are increasingly using water trading and marketing arrangements to improve water use. Despite this progress, the expanded use of formal water trading and marketing arrangements remains highly contentious. The complexity of water trade in wider social-ecological systems also means that trading is not able to comprehensively resolve all socio-economic issues around water use (Meinzen-Dick, 2007). Indeed, Grafton et al. (2016) provide a critical review of the arguments for and against water trading, but conclude that while both social and environmental goals are compatible with water markets, careful design and effective oversight are required for any broader jurisdictional application.

2.2.2   Necessary Conditions

The capacity of water access and allocation arrangements to allow water trading critically depends on local circumstances, the range of future scenarios (and the extent to which they can be known) and the available regulatory architecture (Maestu and Gómez Ramos, 2013). For example, it is common not to control or limit the extraction of water for livestock watering, but access arrangements for cropping can vary enormously by location. Thus, differences in approaches to water trading across locations requires an assessment of what works, and under what conditions (Easter and Huang, 2014).

There are many institutional factors that should be considered when governments contemplate the establishment of water entitlement and allocation regimes. Institutional arrangements necessary to enable efficient and equitable water trading should be in place well before scarcity is realised, to prevent over-allocation. Proactive management of water resources can be achieved by early institutional investments in resource measurement, data collection and governance capacity. Consequently, many of the modifications and enabling conditions required to establish and sustain water markets are also issues associated with sound water resource governance generally (Garrick, 2015).

Matthews (2004) highlights ten questions relevant for any discussion about the establishment or reformation of any water rights system, which have significance for water managers interested in subsequent market adoption. These include: how any rights to water are currently specified, distributed and prioritised; whether existing rights are tradeable in nature, or if transformation would be required; how clear are current operational water use rules, and can they assist/hinder transfers; how certain are we of our data on current source, supply, usage and measurement; how to enforce change or compensate losers in the modifications proposed, and who will achieve this; and are all aspects of the system (for example, groundwater interaction, return flows, losses, and so on) accounted for in the design of water markets. Issues that might help to stimulate water markets are also discussed, such as adopting uniform rights across all uses (but with heterogeneous use-reliability or preferences), increased water pricing, removing spatial limitations to use, and adopting a national registry system (ibid.). This highlights the need to design water rights and, if necessary, to respecify them as part of an integrated reform agenda. When changes are required such that water trading efficiently and equitably retains water use within sustainable limits, then there may also be a need for institutional capacity-building and adaptive governance arrangements to ensure effective and sustainable implementation (Garrick, 2015; Marino and Kemper, 1998).

The literature and practical experiences show that water markets are far from a simple panacea for water reallocation problems. Rather, they are often one of the more complex economic instruments to design, develop, implement and sustain over time. Based on the Australian experience of water markets, Young (2014a) offers six valuable institutional design principles: (1) separate water access arrangements into their various component parts; (2) assign any policy instruments for specific purposes only, and do not use multi-instruments; (3) design instruments with hydrological integrity; (4) keep transaction costs as low as possible; (5) assign risk to one interest group; and (6) ensure robustness of a system through proper accounting for water uses. Although there remain relatively few examples of water markets around the world, and expected benefits from marketing are often unmet due to complex impediments, analyses of global water markets suggest that other essential prerequisites exist. These include: initial allocation transparency; legal clarity and certainty; administrative capacity to cope with changing use arrangements; and vertically and horizontally nested arrangements intended to keep institution costs as low as possible (Grafton et al., 2010). Further, Perry (2013) proposes an ABCD+F (accounting, bargaining, codification, delegation and feedback) list of requirements for effective water resource management. Unfortunately, inadequate institutional capabilities of many countries mean that the journey to water trading arrangements that adequately respond to water security will be long and arduous (Grafton et al., 2016), and firmly out of the reach of some developing nations until such issues are resolved.

The Organisation for Economic Co-operation and Development (OECD) (2015), drawing upon Young (2013), developed a 14-point 'health' checklist for water resource allocation institutional design. Trading arrangements are last in this checklist, suggesting that major transformational reform may be initially necessary together with careful attention to sequencing, to avoid lock-in arrangements that make further transition to water trading arrangements politically difficult. Indeed, we argue that ongoing debate about the merits, or otherwise, of water trading continues because the institutional arrangements used to manage water resources have not been adequately designed to manage water scarcity. As a result, naïve decisions to allow 'unfettered' water trade prior to the reconfiguration of the administrative arrangements to adequately manage water supply and demand can be damaging to, rather than support-ive of, water security (Maestu, 2013; Young, 2014b). Moreover, immature governance arrangements have led to calls for the adoption of non-market approaches to reallocation in regions where water rights are poorly defined and/or institutional capacity is limited (Marston and Cai, 2016).

Alternatives to water markets exist, but may result in less sustainable and effective outcomes in the longer term. Where environmental watering objectives increasingly feature in policy-making, market-based reallocation approaches offer attractive and practical means for future adaptation in the face of uncertainty inasmuch as they offer management discretion over water use. Like Garrick et al. (2009), we emphasise that additional enablers such as necessary administrative procedures, organisational development/capacity to affect transfers, and adaptive mechanisms to overcome legal, economic, cultural and environmental barriers are required.

We contend that a desire or need for marketing reallocation arrangements will grow naturally in many contexts from the adoption of administrative or collective reallocation arrangements, prompting an increasing practical requirement for a water market framework. The numerous and complex barri-ers to water reallocation raised by Marston and Cai (2016) motivate the need for a water market framework that can be used by water managers and planners around the world. Such a common non-prescriptive framework to evaluate the appropriateness of water allocation arrangements to facilitate low-cost trading, and how they might be developed in differing contexts, has yet to be produced (Grafton et al., 2016). Our purpose here, therefore, is to provide a very first framework attempt to fill this important knowledge and practise gap.

2.3　DEVELOPING WATER MARKETS FURTHER

2.3.1　Exploring the Case of the Most Advanced Water Market in the World: the MDB

While we acknowledge the limits on transferability arising from contextual circumstances that allowed markets to flourish in Australia, lessons drawn from a jurisdiction with long-running and successful water rights/market arrangements may offer a basis for such a framework. In particular, the south-ern Murray–Darling Basin (MDB) trading arrangements are often held up as a model for the rest of world to follow (Perry, 2013). The MDB comprises four Australian states and one territory: Queensland (QLD), New South Wales (NSW), South Australia (SA), Victoria (VIC) and the Australian Capital

Territory (ACT). This system is federally managed under joint agreement. An independent Authority is responsible for Basin-wide planning, with states responsible for the issuing of entitlements and management of water use within agreed limits. Federal responsibilities, primarily through the Authority, include: setting, monitoring and enforcing water market rules; monitoring, evaluation and enforcement of the Basin Plan enacted in 2012; determining water allocations for the environment; and prioritising annual environmental watering (Hart, 2015).

Within-state water allocation trades (that is, spot or temporary trade) have been occurring since 1983 in NSW and SA, and since 1987 in Victoria. Water entitlement (that is, permanent) trades have been allowed since 1983 (SA), 1989 (NSW, QLD) and 1991 (VIC); with interstate trades possible since 1995 (Wheeler et al., 2014a). Water allocation trade has grown substantially in the MDB since agreements to unbundle water licences from land (1994) and the introduction of a cap (1995) on further surface water extractions and use. Agricultural producers, by far the biggest MDB water users (ABS, 2013), have become more accepting of water markets over time and found it beneficial to their business (Grafton et al., 2016). Reviews of the economic impact of water trading in the MDB have found it increased regional domestic product by AU$4.3 billion during the last major drought (2006–11) (NWC, 2012). Further, between 2000–2001 and 2007–08, despite a 70 per cent decline in MDB irrigated surface water, water trade, changes in farm crop prices and other adaptations meant that the adjusted gross value of irrigated production fell only 20 per cent (Kirby et al., 2014). Water trading is now widely used as a risk management strategy (Nauges et al., 2016; Zuo et al., 2014), has been of considerable social and economic value to individual water users and to rural communities, and has resulted in positive environmental outcomes (NWC, 2012).

Expected negative trade impacts, such as reductions in regional spending, employment and public services as a consequence of permanently traded water out of districts via markets (Alston and Whittenbury, 2011) and stranded infrastructure assets, have largely been avoided in the MDB. Notwithstanding the successes of water trading in the MDB, there are deficiencies or gaps that currently exist in the MDB which have arisen from current and historical policy, that include: unnecessary trade barriers, the need for improved water market and weather information, limited types of water trade products, inadequate understanding of return flow impacts, and possible future lock-in of some enterprises, such as perennial production systems (Grafton et al., 2016). Nevertheless, the southern MDB offers a valuable context from which to draw insights that assist in the development of a water market readiness assessment framework (Box 2.1).

BOX 2.1   INSIGHTS FROM THE AUSTRALIAN EXPERIENCE IN THE DEVELOPMENT OF WATER TRADING AND MARKETING ARRANGEMENTS

1. The legacy of prior licensing decisions can result in markets causing over-allocation problems to emerge in a manner that erodes the health of rivers, aquifers and ecosystems.
2. Transaction and administrative costs are lower when entitlements are defined using a unit share structure, and not as an entitlement to a volume of water.
3. Market efficiency is improved by using separate structures to define entitlements, manage allocations and control the use of water.
4. Early attention to the development of accurate licence registers is critical and a necessary precondition to the development of low-cost entitlement trading systems.
5. Unless water market and allocation procedures allow unused water to be carried forward from year to year, trading may increase the severity of droughts.
6. Early installation of meters and conversion from area-based licences to a volumetric management system are necessary precursors to the development of low-cost allocation trading systems.
7. Difficulties will be encountered within communities to plan for (and believe in) an adverse climate shift, but water-sharing plans that account for a climatic shift to a drier regime must be developed.
8. The allocation regime for the provision of water necessary to maintain minimum flows, provide for conveyance and cover evaporative losses need to be more secure than that used to allocate water for environmental and other purposes.
9. Unless all forms of water use are accounted for, entitlement reliability will be eroded by expansion of unmetered uses such as plantation forestry and farm dam development, increases in irrigation efficiency, and so on.
10. Unless connected groundwater and surface water systems are managed as an integrated resource, groundwater development and substitution will impact on the future allocation (and use) of surface water.
11. Water use and investment will be more efficient if all users are exposed to at least the full lower-bound cost (preferably the upper-bound cost) of water supply. One way of achieving this outcome is to transfer ownership of the supply system to these users.

12. Manage environmental externalities using separate instruments so that the costs of avoiding them are reflected in the costs of production and use, in a manner that encourages water users to avoid creating them.
13. Remove administrative impediments to inter-regional trade and inter-state trade.
14. Markets will be more efficient and the volume of trade greater if entitlements are allocated to individual users rather than to irrigator-controlled water supply companies and co-operatives.
15. Equity and fairness principles require careful attention to and discipline in the way that allocation decisions and policy changes are announced.
16. Water markets are more effective when information about the prices being paid and offered is made available to all participants in a timely manner.
17. Develop broking industry and avoid government involvement in the provision of water brokering services.

Source: Adapted from Young (2010).

2.3.2 Developing Fundamental Water Market Enablers

The first step toward developing a water market readiness assessment framework involved establishing a set of prerequisites and fundamental water market enablers, as derived from our literature review and other countries' water market experiences or knowledge. Table 2.1 summarises the fundamental issues to be considered, and provides some key examples of the questions that could be used by practitioners to evaluate the need for, and development paths toward, water markets. The resultant set of prerequisites have been transformed into a series of water market enabling and constraining factors; with the full set of relevant questions provided in the Appendix. The Appendix also provides further detail of prioritisation and key importance of specific issues, with five stars indicating high priority and importance for water trade to be successful, and one star indicating low priority/importance.

2.4 A WATER MARKET READINESS ASSESSMENT (WMRA) FRAMEWORK

2.4.1 Developing the Framework

After further refinement and discussion of the insights drawn from our literature review and knowledge, and considering issues relevant to developing countries with low institutional capacity, a conceptual water market readiness

Table 2.1   *Identifying fundamental water market enablers*

Fundamental issues	Key example questions to guide discussion/thinking
Property rights/institutions: Unbundled, individuals versus environment, risk assignment, adaptive, etc.	Does legislation exist which gives a clear understanding of rights to water for individuals/corporations and other legal entities? If so, is the degree of attenuation clear, and which legislation (or pieces of legislation) are pertinent? Does institutional capacity exist in the country to allow robust, transparent and secure water reform?
Governance: Legislation, water sharing plans, information availability, water allocation announcements, compliance, etc.	Are enabling resources (such as information, planning resources and registers) available, reliable, legitimate and trustworthy? Is the administrative culture and behaviour of those involved in making decisions respected and trusted?
Hydrology: Connected systems, salinity and water quality considerations, limit and consequences of breach → environment → end of system, do we know what we don't know etc.	Is the hydrology of the system well understood, well documented, monitored and reported on in a way that is supportive of trade and is sympathetic to: The resource constraint, and The extent to which the knowledge of the resource is complete?
Entitlement registers and accounting systems: Ownership, trading rules, tracking use	Does the supplier have the systems, resources and technology to monitor use, and to ensure use is within constraints, licence/entitlement conditions? Are the registers and accounting systems used to track and enforce compliance robust?
System type: Regulated/unregulated, surface water/ groundwater, connectivity, etc.	What is that status of infrastructure and what are the costs of accessing water in the system, and at various parts of the system?

Fundamental issues	Key example questions to guide discussion/thinking
Adjustment:	
Heterogeneity → Gains from trade, societal pressures, early-mover advantage, etc.	Is there a sufficiently diverse (potential) market for water use in the system so as to facilitate trade (willing buyers and sellers with different use profiles in terms of value added per $ of water) and what is the likely magnitude of these gains (ex-transaction costs)?
Externalities	Are effective arrangements in place to maintain water quality, ensure environmental outcomes, facilitate navigation, hydro-power generation, etc.?

assessment (WMRA) framework that involved three key steps was developed (Figure 2.1).

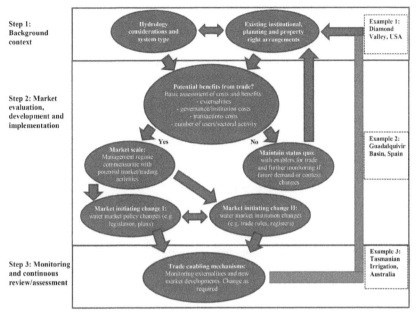

Source: Wheeler et al. (2017).

Figure 2.1   *Conceptual assessment approach for considering the readiness of jurisdictions for water markets*

The first step of the WMRA ('Background context') is a scoping exercise that establishes the context of a proposed market such as planning considerations, and established resource knowledge that allows for a definitive cap or an initial allocation/extraction level. The importance of institutional capacity is critical in the first step, including careful consideration of: broad reviews of the status and maturity of water rights, governance and institutional capacity, the current level of infrastructure development and operational rules, and the availability and quality of water data. Where these are insufficiently progressed and do not meet minimum requirements (for example, those stipulated in Matthews, 2004), non-market reallocation arrangements may be more appropriate and further use of the framework should cease until further institutional capacity is developed. However, if deemed to be sufficient for market-based arrangements, then further steps may be taken.

The second step ('Market evaluation, development and implementation') goes beyond simply considering capacity, and assesses the current institutional arrangements that support or impede trading (discussed further later on). The third step ('Monitoring and continuous review/assessment') outlines a continual review and assessment of water trading in the pursuit of further gains as experience emerges. The importance of considering sequential steps for the successful development and implementation is noted by Young (2014b). Our approach is consistent with Young's sequential market steps, but what we propose is non-prescriptive, and therefore not necessarily sequential or mutually exclusive. That is, although the process may appear linear, in fact progress can be made in parallel or by entering the framework at any relevant stage where enabling conditions permit. Our three steps thus provide a logical order for assessing the necessary preconditions where they may not currently exist, and offer multiple entry/exit points for water managers/planners. A more detailed analysis of each step follows.

2.4.2　Step 1: Background Context

The first step involves an evaluation of current institutional, legislative, planning and regulatory capacity to facilitate and/or allow water trade ('Existing institutional, planning and property right arrangements'). We expect that for many countries, institutional capacity is insufficiently progressed and does not meet minimum requirements to be able to proceed with water market reforms. Effective water trading arrangements fundamentally require a clearly specified set of entitlement and allocation arrangements that are monitored and enforced. Arguably one of the first administrative challenges is to establish a regime that ensures that water users can understand their entitlements and how to transfer them. This is challenging if the current situation involves significant over-extraction/consumption. Ideally, the water governance arrangements should: (1) fully specify each share of the resource in perpetuity while allowing for changes in the proportion allocated to each share; (2) define the opportunity to use water in an unambiguous manner that fully assigns responsibility for managing supply risk to users; (3) be enforceable and ensure that the only way one user can access more water is to convince someone else to use less water; and (4) keep transaction costs as low as possible.

Water scarcity is a common motive for implementing water markets (Easter et al., 1998). Thus, in addition to institutional capacity issues, a clear calculation and definition of the total resource pool available for consumptive use, and how that may change over time ('Hydrology considerations and system type'), needs to be conducted (Freebairn, 2005). This should be combined with clear rules for the allocation of that defined resource pool.

2.4.3   Step 2: Evaluation, Development and Implementation

Effective water market governance includes the separation of regulatory, policy, commercial and operational functions associated with the water resource(s) in question. It also includes clear and consistent trading rules, as well as compliance and enforcement mechanisms (NWC, 2011a). There must also be a sufficient volume of exchange to overcome potential inefficiencies. Hence, the second step of the framework is to assess the 'Potential benefits of trade'. For example, there has to be consideration of the number of individuals involved, the homogeneity of water use, the potential benefit from trade and changes in water use behaviour that can be derived, the direct costs associated with the governmental and institutional policy reforms necessary to enable trade, and the transaction costs associated with implementation and ongoing use. When the results suggest that there are significant net benefits, then this information needs to be packaged into two narratives: one that can be understood by the public, and a second much more comprehensive narrative for consideration by professional analysts and those likely to be involved in facilitating implementation (Young, 2016).

Other reallocation inhibitors include high trade and transaction costs (for example, where participants are charged very high costs to transfer entitlements or allocations, or face extremely long processing times to do so). There may also be price-setting influences as a result of limited competition among market participants, and potential externalities where the buyers and sellers do not exclusively enjoy and incur all of the benefits and costs (respectively) associated with trade. If a critical trading volume and benefits can be achieved, there may be scale economies associated with broader market implementation. This involves a move to 'Market initiating change I' in Figure 2.1, where broader legislation is put in place to allow for water reform, which then leads on to 'Market initiating change II' where more detailed water reform leads to improvements like registers and clearer trade rules. If the assessments at Step 2 points to modest gains from trade, then the best policy for water managers/ planners may be to 'Maintain status quo', but at the same time to instigate enablers for trade (which would allow transfers between a small number of individuals but not a fully functioning water market per se) back at Step 1's 'Existing institutional, planning and property right arrangements'. Continual monitoring is also needed to assess future demand or context changes (which may move the planner to the sub-step of 'Market initiating change I').

2.4.4   Step 3: Monitoring and Review

The development of water trading typically requires continuous monitoring, review and assessment whatever the stage of market development; but especially where it is to be sustained over the long term. This mainly involves further development of the 'Trade enabling mechanisms' in Step 3. This will include efforts to limit/reduce transaction costs, adapt to new information as it arises, and to scan for unanticipated externalities and opportunities for refined market products (for example, option contracts or water banks) (Wheeler et al., 2013; 2014a; 2014b). Such monitoring may, in turn, reveal additional changes to information sources or collection methods that then require legislative change, or new planning requirements/infrastructure projects to improve trade capacity (putting the planner back to review Step 1's 'Existing institutional, planning and property right arrangements').

Experimentation and adaptation is not uncommon in countries where water markets have been previously implemented (Maestu, 2013), and building flexibility into trading rules and procedures serves to enhance water security and management robustness, especially where risk needs to be accommodated along with future uncertainty (Garrick and Hall, 2014). It is also possible that flexibility in trading arrangements may provide opportunities for political intervention in the market. Thus, an appropriate balance between rule and process adaptation and surety is highly desirable. In the next section we explore applications of the framework in three examples to assess its applicability in different circumstances.

2.5　APPLIED EXAMPLES

Three location-specific examples to apply the framework were chosen based on several considerations. First, we focused on jurisdictions where water markets are discussed as a possible means for responding to water security. Second, the particular regions reflect areas where water stress by 2040 as a ratio of withdrawals to supply is expected to be high (that is, 40–80 per cent; Maddocks et al., 2015). Third, we chose countries at different levels of water market development, including: incomplete unbundling of land and water rights; uncertainty about water right definitions; unfinished catchment or water planning processes; limited time for development of trading rules; reluctance to cap extraction levels; an absence of water entitlement registers, water allocation accounts, inadequate metering, trading platforms, trade processing systems and market information; and/or a lack of administrative experience.

2.5.1　Applied Example One: Nevada's Diamond Valley

The first example is the Diamond Valley area (see Figure 2.2) near Eureka in central Nevada, United States of America (USA). This location has a tightly connected and rapidly depleting groundwater resource, with no connections to other water resources. In addition, the community of water users is small and all the actors know one another.

Source: Wheeler et al. (2017).

Figure 2.2   Diamond Valley area

Applying Step 1's 'Hydrology considerations and system type' criteria, the Diamond Valley area contains a rich groundwater aquifer servicing irrigation, urban, mining and livestock uses. Current groundwater extraction rights of 160 354 ML have been issued, with annual use estimated around 114 715 ML, while the sustainable annual yield is estimated at just 43 170 ML (hence indicating that around 70 per cent of rights need to be retired to ensure sustainable water use). This over-use has resulted in aquifer declines of about 1–4 feet per year. Full aquifer depletion is expected in 30 years. Approximately 95 per cent of committed water rights in the Valley are vested with primary or secondary irrigation farms. Initial plans to achieve a reduction in rights focused on payments to farmers who agreed to retire their rights, based on acreage and economic value calculations. No consistent or agreed value for the water was available, because no market existed. The estimated total cost of this programme was US$45 million over 50 years, dependent on assumed linear reductions over that period and farmers' willingness to participate (Hansford, 2014).

In terms of Step 1's 'Existing institutional, planning and property right arrangements', water entitlements (rights) in the Valley are issued under the prior appropriation system where the oldest licences are assigned full allocations ahead of more recently issued (junior) rights. Also, under a current beneficiary-use requirement there is no incentive to innovate and ensure efficient water use. If all the water assigned to a user is not used, then they risk the curtailment of that right, which incentivises over-consumption. Given the demand for extraction change and reductions in over-use, in 2015 the State Engineer decreed that all claimants to water rights (surface and groundwater) would need to provide proof to substantiate existing rights and that the community would have ten years to prepare an approved groundwater management plan. If they failed to do so, the State Engineer would curtail water use on the basis of seniority.

The threat of water regulation has helped in identifying that there are net benefits from trade and moving towards different reallocation systems. In terms of assessing Step 2 of the framework ('Potential benefits from trade'), commissioned studies in the Valley show evidence of significant gains from trade versus curtailment (e.g., Zeff et al., 2016). This reality has provided the impetus to undertake the legal and institutional reforms necessary to make sure all existing property rights and institutions are now conducive to trade. Consequently, the Diamond Valley plan is now well advanced, and in 2017 proposed a five-year process to convert all existing rights into tradeable shares; make annual allocations in proportion to the number of shares held; meter and enforce (with penalties) water use; gradually curtail extraction to a sustainable level cap over a 30-year period; create a State Guarantee of share register and water account integrity; and implement efficient short-term trade; and management via a local governance board (Zeff et al., 2016).

The political and community acceptance of such a radical change in water governance was able to be achieved by allowing reversibility in the plan. For example, if at the end of five years the community believed the new sharing regime to be inferior to the original priority regime, then all new shares would be cancelled and the old regime resumed (which is not expected). Summarising all this information in Table 2.2 (in the Summary section, below), and applying the WMRA framework, suggests that the Diamond Valley is at the early stages of Step 2, with a range of market-initiating changes still required to achieve market development.

2.5.2 Applied Example Two: Guadalquivir River Basin, Spain

Spain has often been identified in the literature as the blueprint for future water markets in the European Union (Hernández-Mora and del Moral, 2015). In particular, the Guadalquivir River Basin (GRB) in Spain is an interesting applied example as it is expected to play an important role in the country's water market development and expansion. The GRB has the country's longest river (Guadalquivir River, 666 km in length), spanning 12 independent provinces and four autonomous irrigation communities/regional governments.

Demand is dominated by irrigated agriculture (89 per cent used for both annual and perennial crops). Surface water resources provide the bulk of supply (74 per cent), supplemented by groundwater (Bhat and Blomquist, 2004). Total consumptive water demand in the Basin is 3 845 100 ML per annum, while estimated average water resources are around 3 607 600 ML/year. Hence, water scarcity is a defining feature of the GRB, which has historically been addressed through the construction of 65 interconnected dams and storages. There is no cap or closed basin arrangement, and new water concessions (rights) can still be granted and new irrigated zones planned. Consumptive water demand is dependent upon seasonal water availability and proportional sharing rules based on priority classes (which meets some of the requirements in 'Hydrology considerations and system type' in Step 1).

In terms of addressing key fundamentals of 'Existing institutional, planning and property right arrangements' in Step 1, irrigation water rights date back to the 1879 Water Act, updated in 1985 to provide renewable legal rights for up to 75 years (which are still attached to land and managed by a Community of Irrigators under an 'administrative concession'). The need to address water scarcity and the benefits of potential water trade motivated Law 46 in 1999, resulting in the introduction of temporary trade markets between Irrigation Community members on a limited scale (under requisite regulatory limits around use-type, location, extraction conditions and/or return flow requirements). Permanent sales of water without land are not legally possible and/or difficult to ratify through the collective-management arrangements. However, a significant impediment to trade has arisen from some historical decisions taken by the River Basin Authority, which have undermined the trust of some irrigation communities with regard to the security of their property rights. Uncertain right arrangements have thus curtailed involvement in larger-scale GRB trade, and reduced the benefits associated with transfers between hydrologically interconnected irrigation communities.

In terms of assessing the 'Potential benefits from trade' in Step 2, while water charges and scarcity within communities have driven temporary trade and water bank trades since 1999, the scale of market trade remains low due. Other factors include: (1) agricultural user scepticism toward trading and socio-economic concerns about trade consequences; (2) irrigator substitution of surface water for other sources such as groundwater (although laws have been implemented to limit such substitutions, approximately 80 per cent of groundwater users have not registered for a permit); (3) high transaction costs (lengthy times to assess and approve lease transfers, coupled with many conditional restriction requirements); and (4) lack of information and specialised trade platforms and/or third-party providers (Hernández-Mora and del Moral, 2015). However, the GRB's second water management Basin Plan (2015–21) plans to review such water right issues, and may implement a range of further property right and institutional reforms. From the WMRA framework assessment it is clear that that despite water market objectives this applied example remains situated in a Step 2 loop, sitting between 'Market initiating change I/ II' and 'Maintaining status quo'.

Although considerable effort has been put into trade-enabling mechanisms and arrangements, these are still not supported by separation of land and water property rights (institutional change) and broader perceptions of property right security. In sum, the GRB is an example of a jurisdiction that is somewhat advanced in its water market development/implementation, but which has fundamental issues with governance impartiality and trustworthy systems. Before movement can be made towards permanent trade, the following issues need addressing: capping all GRB extraction (including groundwater) to be consistent with the European Union's Water Framework Directive; separating land and water rights; and investments in property right security and institutional reforms.

2.5.3   Applied Example Three: Tasmanian Irrigation, Australia

The third example is in Australia, but outside the MDB. Tasmania is an island state to the south of the mainland with catchments that are hydrologically disconnected, creating independent allocation planning areas, diverse management arrangements and differing potential for water market development. The state has a cool temperate climate that can vary widely regionally between cold and hot extremes. Much of the state's irrigation districts are located on the eastern side of the island in regions where water capture, storage and delivery is possible.

In recent years, many new irrigation schemes have been proposed with the backing of state and federal grants to further economic development (as outlined by the Tasmanian Water Development Plan). This development objective is being achieved through public–private partnership arrangements (similar to past irrigation subsidy proposals experienced on mainland Australia) with a budget of AU$220 million (Tasmanian Irrigation, 2012). Many catchments are not fully allocated, in the sense that water available for extraction/consumption exceeds current water entitlements. The exception is where hydro-electricity generation is present (NWC, 2013a).

In terms of assessing Step 1's key fundamentals of 'Existing institutional, planning and property right arrangements' and 'Hydrology considerations and system type', Tasmania's state agencies are well advanced. They have developed surface water and groundwater hydrological models, coupled with a freshwater ecosystem value database. This information provides sustainable yield and extraction catchment data, stress rankings and a management interface capacity (Tas DPIPWE, 2017). In addition, Tasmanian Irrigation (TI) used historical hydrology data tested against climate change runoff models developed by the Commonwealth Scientific and Industrial Research Organisation (CSIRO) sustainable yields project (CSIRO, 2009) to help set scientifically defined sustainable consumptive limits for each irrigation scheme (NWC, 2013b). Mandatory farm water access plans are used to ensure that land is managed in accordance with the water development plan. Irrigation schemes have a projected 100-year life, and are planned to supply water at an average annual reliability of 95 per cent.

Tasmanian planners are currently focused on developing and supporting the institutions needed for water trade among affected stakeholders, particularly those in the agricultural sector. Water entitlement and allocation arrangements are being designed to facilitate water trading, as it was assessed that there were considerable 'Potential benefits from trade' (Step 2) because of the number of irrigators, the heterogeneity of agricultural production in the region, and future demand for Tasmanian agricultural production (especially given the likely pattern of climate change across Australia). For example, many viticultural producers are relocating to Tasmania as they are unable to continue growing key varieties in traditional production areas on the Australian mainland given rising temperatures, changing seasons and increased pest burdens.

State planners have also made significant progress towards legislative changes, that is, 'Market initiating change I/II' in Step 2. For example, as each irrigation region is developed, unbundled water entitlements will be sold via an open tender using a reserve price. Moreover, ongoing operational costs, including asset refurbishment or renewal provisions, will not be subsidised and must be met by annual water charges on licence holders (making them consistent with the National Water Initiative). Other market reforms include a water register that defines licence ownership, protects registered financial interests and facilitates both temporary and permanent transfers. An online trading platform has also been established to reduce trade transaction costs; although an identified institutional weakness is that the TI register does not have the capacity or the legal obligation to record price data (NWC, 2013a).

The application of the WMRA framework (Table 2.2) to Tasmania indicates that many prerequisites for effective and efficient trade are present.

For example, there is a cap on resources; a water plan and rigorous planning process; robust entitlement registers and water accounting arrangements with all barriers to trade (other than those hydrologically justified) removed. In addition, there are no restrictions on ownership and no requirements to use all allocated water, and any unintended trade-related environmental externalities are adequately managed by regulation.

TI has therefore reached Step 3 of the framework, 'Trade enabling mechanisms', where change and adaption is needed as a result of trade enabling mechanisms. Further development of trade between users and monitoring of progress will help determine future changes or reforms. As a consequence, there may be a need for further adjustments to the steps of 'Existing institutional, planning and property right arrangements and 'Market initiating change II', such as improved data collection and other water planning arrangements needed to sustain market development.

2.6   SUMMARY AND DISCUSSION OF KEY INSIGHTS FROM THE REGIONAL EXAMPLES

Table 2.2 provides the overall summary, and the application of the framework to our specific regional examples. Apart from the key fundamental need for strong, trusted institutions and governance to allow the development of water markets in any jurisdiction, which limits the regions where water trade can be successfully applied, there are three important lessons that emerge from the application of the WMRA framework to the examples: (1) unbundling; (2) sequencing; and (3) 'never waste a good crisis'.

In many jurisdictions, unique bundling arrangements are used to protect the resource by keeping use within sustainable limits. A water licence may, for example, require that the water be used only for a specific purpose and applied in a specific manner. Such arrangements discourage conservation and make it difficult for administrators to reduce total use. In Spain, the continued existence of uncertain property right enforcement limits trading outside a communal area, is an important barrier to efficient water use and discourages investment; which is avoided in the MDB, given the fact that entitlements are issued to users rather than the community and local controls on the trade of shares and/or annual allocations confined to the setting of reasonable exit fees. The costs of trading shares and allocations are low and can be finalised quickly. This avoids any obligation to use all annually allocated water and incentivises investment in water conservation; especially when it is possible to save water for subsequent use, as will soon be the case in the Diamond Valley. The Tasmanian example highlights the value (and necessity) of unbundling water rights ahead of introducing trade arrangements. These arrangements encourage the management of long-term supply risk with water via water

Table 2.2   *WMRA application summary*

Key fundamental market assessors	Diamond Valley	Guadalquivir	Tasmania
Property rights/institutions			
1. Water legislation	✓	✓	✓
2. Unbundled rights	X	X	✓
3. Rights transferable	X	✓	✓
4. Rights enforceable	✓	X	✓
5. Constraints between connected systems	✓	X	✓
Hydrology			
1. Documented hydrology system	✓	✓	✓
2. Understanding of connected systems	✓	✓	✓
3. Future impacts modelled	✓	✓	✓
4. Trade impacts understood	✓	✓	✓
5. Resource constraints understood	✓	X	✓
6. Resource constraints enforced (e.g., existence of a cap/closed basin)	X	X	✓
Externalities/governance			
1. Strong governance impartiality	✓	X	✓
2. Existence of externalities understood	✓	✓	✓
3. Water use monitored	X	✓	✓
4. Water use enforced	X	✓	✓
System type			
1. Suitability of water sources for trade	✓	✓	✓
2. Transfer infrastructure availability/suitability	✓	✓	✓
3. Regulation requirements for trade	X	✓	✓
Adjustment			
1. Gains from trade (no. users/transaction costs/ diversity of use)	✓	✓	✓
2. Political acceptability of trade	✓	X	✓
Entitlement registers and accounting			
1. Trustworthy systems	X	X	✓
2. Trade and market information availability	X	X	✓
Trade step reached	Steps 1–2	Step 2	Step 3

Note: X indicates further reform is required for that issue in the particular regional example; ✓ indicates that there is good evidence supporting that particular part of the assessment; while a smaller ✓ indicates that there is positive but still limited evidence, and thus room for improvement.

entitlement trade, while water allocation trades encourage efficient water use on a day-by-day basis.

As identified by Young (2014b), correct sequencing of water reforms is critical. A recognition of the limits to water use, associated restrictions on further extraction and genuine incentives to change behaviour are all essential factors in successful water market adoption. In particular, rights and water accounting rules need to be consistent with hydrological realities. Return flows, for example, should be included in the water markets accounting framework. In the Spanish example, although return flows are often accounted for within an irrigation community, conflicting allocation requirements and hydrological information about environmental flows place the GRB at high risk of desertification due to water over-extraction. Thus, Hernández-Mora and del Moral (2015) argued that a great deal of further institutional development was required before water trade can be increased. This highlights that implementing water trading without fully implementing the required laws, institutional capacity and administrative systems in a basin will only increase transaction costs; and fuel scepticism about market benefits.

Each of the applied examples provides key lessons about the opportunities that can be garnered from crises, and using trading as a means to keep water use within limits. Arguably, real gains come from the fact that trading encourages the development of administrative systems that are self-enforcing. In Tasmania, the consequences of costly climate change on traditional agricultural production (especially viticulture), the success of MDB water markets (and the ability to consider the lessons from earlier mainland water market establishment) enabled planners to select the most suitable institutional reforms and systems needed. By contrast, in the Diamond Valley the water crisis has resulted in the willingness to implement considerable institutional and property right change and funding to rectify key hydrological information gaps.

Our application of the WMRA framework to the examples in this study indicates that it provides a useful tool for helping water planners/managers to evaluate the relevant information/conditions needed for water markets and, as a consequence, potentially respond to the fundamental problems of water scarcity. The application of the framework suggested that it may serve as a practical, relatively quick and non-prescriptive means for water managers/ policy-makers in different jurisdictions to assess the appropriateness of emergent or developing water market arrangements.

We stress that the development of successful market reforms depends critically on the existence, impartiality (and security) of water institutions. Our applied examples provide evidence of how jurisdictions undertake their planning process and how, during the development of water allocation plans, those with existing rights to water are required to engage as part of extensive stakeholder consultation. Where stakeholder consultation occurs, this may ensure that knowledge held by these groups is not overlooked, reduce community and political opposition, and potentially lower the cost of (further) modifying extraction or consumption limits (Crase et al., 2011). Further, where a resource is shown to be over-extracted/consumed and new information arises about the unsustainable nature of extractions, reductions in use can then be managed in a manner that does not necessarily negatively impact upon market confidence. This is encapsulated in our WMRA framework. Nevertheless, we acknowledge that this framework is only a first step forward; further testing, application and refinement will obviously be required. Further research would help to operationalise WMRA, as would consideration of comparable measures within and across examples/scales and over time, particularly in regard to trading activity.

2.7   CONCLUSION

Water demand management strategies will need to be implemented across the world as regions increasingly grapple with water scarcity. One possible strategy includes the establishment of water entitlement and allocation systems that make rapid, low-cost water trading possible. The applicability of such systems to various regions is often unknown, and there is a dearth of information, guides or manuals that show water managers and planners how to proceed. To assist this process, a WMRA framework (and associated set of questions) was developed to offer practitioners a non-prescriptive three-step framework: (1) assess hydrological and institutional needs; (2) evaluate market, development and implementation factors; and then (3) monitor and continuously assess effectiveness.

The framework was applied to three examples from different countries to help evaluate the potential for water markets to address water scarcity issues, by an assessment of market enabling/constraining conditions. Our preliminary findings suggest that WMRA may help practitioners to identify the reforms needed to help improve existing arrangements, or correspondingly identify that market arrangements are not possible for their region. As with any proposed framework, further testing and application is required to assess its applicability and value.

ACKNOWLEDGEMENTS

None of the authors' affiliations or employment positions provides any conflicts of interest that must be raised. This work was supported by the Australian Research Council [FF140100733, DE150100328 and DP140103946]. The authors gratefully acknowledge the Australian National Water Commission's role in originally driving this research, and for allowing the authors to access material under its care. We are also indebted to helpful comments received from reviewers and to Sara Palomo-Hierro for her help with Spanish water market literature.

PUBLISHER'S NOTE

NOTE

1. Originally published as: Wheeler, S., Loch, A., Crase, L., Young, M., and Grafton, R. (2017). Developing a water market readiness assessment framework. *Journal of Hydrology*, 552, 807–820. https://doi.org/10.1016/j.jhydrol.2017.07 .010. Republished here with some minor changes under the terms of: https:// creativecommons.org/licenses/by/4.0/.

REFERENCES

ABS (2013). *Water Use on Australian Farms, 2011–12*. Australian Bureau of Statistics, Canberra.

Alston, M., and Whittenbury, K. (2011). Climate change and water policy in Australia's irrigation areas: a lost opportunity for a partnership model of governance. *Environmental Politics*, *20*(6), 899–917.

Bakker, K. (2005). Neoliberalizing nature? Market environmentalism in water supply in England and Wales. *Annals of the American Associaton of Geographers*, *95*(3), 542–565.

Bakker, K. (2007). The 'commons' versus the 'commodity': alter-globalization, anti-privatization and the human right to water in the Global South. *Antipode*, *39*(3), 430–455.

Barlow, M., and Clarke, T. (2002). Who owns water? *Nation*, *2*(9), 2002.

Bauer, C.J. (2004). Results of Chilean water markets: empirical research since 1990. *Water Resources Research*, *40*(9), W09S06.

Bhat, A., and Blomquist, W. (2004). Policy, politics, and water management in the Guadalquivir River Basin, Spain. *Water Resources Research*, *40*(8), 1–12.

Chong, H., and Sunding, D. (2006). Water markets and trading. *Annual Review of Environment and Resources*, *31*, 239–264.

Crase, L., and O'Keefe, S. (2009). The paradox of national water savings: a critique of Water for the Future. *Agenda*, *16*(1), 45–60.

Crase, L., O'Keefe, S., and Dollery, B. (2011). Presumptions of linearity and faith in the power of centralised decision-making: two challenges to the efficient management of environmental water in Australia. Western Economic Association International Conference, Brisbane, Queensland, 26–29 April.

CSIRO (2009). *The Science of Tackling Climate Change*, CSIRO General Publication, Melbourne.

Dinar, A., and Saleth, R. (2005). Can water institutions be cured? A water institutions health index. *Water Supply*, *5*(6), 17–40.

Easter, K., and Huang, Q. (2014). Water markets: how do we expand their use? In K. Easter and Q. Huang (eds), *Water Markets for the 21st Century* (pp. 1–9). Springer.

Easter, K., Rosegrant, M., and Dinar, A. (1998). *Markets for Water: Potential and Performance*. Natural Resource Management and Policy series. Kluwer Academic Publishers.

Easter, K., Rosegrant, M., and Dinar, A. (1999). Formal and informal markets for water: institutions, performance, and constraints. *World Bank Research Observer*, *14*(1), 99–116.

Freebairn, J. (2005). Principles and issues for effective Australian water markets. In J. Bennett (ed.), *The Evolution of Markets for Water: Theory and Practice in Australia* (pp. 8–23). Edward Elgar Publishing.

Garrick, D. (2015). *Water Allocation in Rivers under Pressure*. Edward Elgar Publishing.

Garrick, D., and Hall, J. (2014). Water security and society: risks, metrics, and pathways. *Annual Review of Environment and Resources*, *39*, 611–639.

Garrick, D., Siebentritt, M.A., Aylward, B., Bauer, C.J., and Purkey, A. (2009). Water markets and freshwater ecosystem services: policy reform and implementation in the Columbia and Murray–Darling Basins. *Ecological Economics*, *69*(2), 366–379.

Garrick, D., Whitten, S., and Coggan, A. (2013). Understanding the evolution and performance of market-based water allocation reforms: a transaction costs analysis framework. *Ecological Economics*, *88*, 185–205.

Goldman, M. (2007). How 'Water for All!' policy became hegemonic: the power of the World Bank and its transnational policy networks. *Geoforum*, *38*(5), 786–800.

Grafton, R.Q., Horne, J., and Wheeler, S.A. (2016). On the marketisation of water: evidence from the Murray–Darling Basin, Australia. *Water Resources Management*, *30*(3), 913–926.

Grafton, R.Q., Landry, C., Libecap, G., McGlennon, S., and O'Brien, R. (2010). An integrated assessment of water markets: Australia, Chile, China, South Africa and the USA. National Bureau of Economic Research, working paper 16203, Cambridge, MA.

Grafton, R.Q., Libecap, G., Edwards, E., O'Brien, R., and Candry, C. (2012). Comparative assessment of water markets: insights from the Murray–Darling Basin of Australia and the Western USA. *Water Policy*, *14*(2), 175–193.

Griffin, R. (2006). *Water Resource Economics: The Analysis of Scarcity, Policy and Projects*. MIT Press.

Hansford, C. (2014). The cost of rectifying over-appropriation of groundwater in the Diamond Valley, Nevada. Water Resources Association Conference, Las Vegas, Nevada.

Hart, B.T. (2015). The Australian Murray–Darling Basin Plan: challenges in its implementation (Part 1). *International Journal of Water Resources Development*, *32*(6), 819–834.

Hernández-Mora, N., and del Moral, L. (2015). Developing markets for water reallocation: revisiting the experience of Spanish water mercantilización. *Geoforum, 62,* 143–145.

Howe, C., Schurmeier, D., and Shaw Jr, W. (1986). Innovative approaches to water allocation: the potential for water markets. *Water Resources Research, 22*(4), 439–445.

Kirby, M., Bark, R., Connor, J., Qureshi, M., and Keyworth, S. (2014). Sustainable irrigation: how did irrigated agriculture in Australia's Murray Darling Basin adapt in the Millennium Drought? *Agricultural Water Management, 145,* 154–162.

Maddocks, A., Young, R., and Reig, P. (2015). *Ranking the World's Most Water-Stressed Countries in 2040.* World Resources Institute.

Maestu, J. (2013). *Water Trading and Global Water Scarcity: International Experiences.* RFF Press.

Maestu, J., and Gómez Ramos, A. (2013). Conclusions and recommendations for implementing water trading. In J. Maestu (ed.), *Water Trading and Global Water Scarcity: International Experiences* (pp. 334–346). RFF Press.

Marino, M., and Kemper, K. (1998). *Institutional Frameworks in Successful Water Markets: Brazil, Spain, and Colorado, USA.* World Bank.

Marston, L., and Cai, X. (2016). An overview of water reallocation and the barriers to its implementation. *Wiley Interdisciplinary Reviews: Water, 3*(5), 658–677.

Matthews, O. (2004). Fundamental questions about water rights and market reallocation. *Water Resources Research, 40,* W09S08.

McCann, L., and Easter, K. (2004). A framework for estimating the transaction costs of alternative mechanisms for water exchange and allocation. *Water Resources Research, 40*(9), W09S09.

Meinzen-Dick, R. (2007). Beyond panaceas in water institutions. *Proceedings of the National Academy of Sciences, 104*(39), 15200.

Meinzen-Dick, R., and Ringler, C. (2008). Water reallocation: drivers, challenges, threats, and solutions for the poor. *Journal of Human Development, 9*(1), 47–64.

Nauges, C., Wheeler, S.A., and Zuo, A. (2016). Elicitation of irrigators' risk preferences from observed behaviour. *Australian Journal of Agricultural Resource Economics, 60*(3), 442–458.

NWC (2011a). *Strengthening Australia's Water Markets.* National Water Commission.

NWC (2011b). *Water Markets in Australia: A Short History.* National Water Commission.

NWC (2012). *Impacts of Water Trading in the Southern Murray–Darling Basin between 2006–07 and 2010–11.* National Water Commission.

NWC (2013a). *Australian Water Markets: Trends and Drivers 2007–08 to 2011–12.* National Water Commission.

NWC (2013b). *Water Management and Pathways to Sustainable Levels of Extraction: Issues Paper.* National Water Commission.

OECD (2015). *Water Resources Allocation: Sharing Risks and Opportunities.* OECD Studies on Water series. OECD Publishing.

Perry, C. (2013). ABCDE+ F: a framework for thinking about water resources management. *Water International, 38*(1), 95–107.

Sadoff, C., Hall, J.W., Grey, D., Aerts, J.C.J.H., Ait-Kadi, M., et al. (2015). *Securing Water, Sustaining Growth: Report of the GWP/OECD Task Force on Water Security and Sustainable Growth.* University of Oxford.

Segerfeldt, F. (2005). *Water for Sale: How Business and the Market Can Resolve the World's Water Crisis.* Cato Institute, Washington, DC.

Shah, T., and Ballabh, V. (1997). Water markets in north Bihar: Six village studies in Muzaffarpur District. *Economic and Political Weekly*, *32*(52), A183–A190.

Tan, P.-L. (2005). A property framework for water markets: the role of law. In J. Bennett (ed.), *The Evolution of Markets for Water: Theory and Practice in Australia* (pp. 56–75). Edward Elgar Publishing.

Tas DPIPWE (2017). *Investing in Irrigation*. Tasmanian Department of Promary Industries, Parks, Water and Environment.

Tasmanian Irrigation (2012). *Just Add Water: An Innovation Strategy for Tasmania*. Tasmanian Irrigation Pty Ltd.

UN Water (2013). *Water Security and the Global Water Agenda: A UN-Water Analytical Brief*. United Nations.

Wheeler, S.A., Garrick, D., Loch, A., and Bjornlund, H. (2013). Evaluating water market products to acquire water for the environment in Australia. *Land Use Policy*, *30*(1), 427–436.

Wheeler, S., Loch, A., Crase, L., Young, M., and Grafton, R. (2017). Developing a water market readiness assessment framework. *Journal of Hydrology*, *552*, 807–820. https://doi.org/10.1016/j.jhydrol.2017.07.010.

Wheeler, S.A., Loch, A., Zuo, A., and Bjornlund, H. (2014a). Reviewing the adoption and impact of water markets in the Murray–Darling Basin, Australia. *Journal of Hydrology*, *518*, 28–41.

Wheeler, S.A., Zuo, A., and Bjornlund, H. (2014b). Australian irrigators' recognition of the need for more environmental water flows and their intentions to donate water allocations. *Journal of Environmental Planning and Management*, *57*(1), 104–122.

WWAP (2014). *The United Nations World Water Development Report: Water and Energy*, World Water Assessment Programme. UNESCO.

Young, M. (2010). *Environmental Effectiveness and Economic Efficiency of Water Use in Agriculture: The Experience of and Lessons from the Australian Water Reform Program*. Consultant report prepared for the OECD, Paris, France.

Young, M. (2013). Improving water entitlement and allocation. Background paper for the OECD project on water resources allocation (unpublished).

Young, M. (2014a). Designing water abstraction regimes for an ever-changing and ever-varying future. *Agricultural Water Management*, *145*, 32–38.

Young, M. (2014b). Trading into trouble? Lessons from Australia's mistakes in water policy reform sequencing. In K. Easter and Q. Huang (eds), *Water Markets for the 21st Century* (pp. 203–214). Springer.

Young, M. (2016). Transformational change. In M. Young and C. Esau (eds), *Environmental and Natural Resource Management: Guidelines for Policy Excellence* (pp. 11–33). Earthscan.

Young, M., and McColl, J. (2009). Double trouble: the importance of accounting for and defining water entitlements consistent with hydrological realities. *Australian Journal of Agricultural Resources Economics*, *53*, 19–35.

Zeff, H., Characklis, G., Kaczan, D., Murray, B., and Locklier, K. (2016). *Benefits, Costs, and Distributional Impacts of a Groundwater Trading Program in the Diamond Valley, Nevada*. Nicholas Institute for Environmental Policy Solutions, North Carolina.

Zuo, A., Nauges, C., and Wheeler S.A. (2014). Farmers' exposure to risk and their temporary water trading. *European Review of Agricultural Economics*, *42*(1), 1–24.

Table 2A.1   *Water market assessment questions*

Issues	Questions to guide discussion/thinking	Priority
Property rights/institutions:		
Legislation	Does legislation exist which gives a clear understanding of rights to water for individuals/corporations and other legal entities? If so, is the degree of attenuation clear, and which legislation (or pieces of legislation) are pertinent?	*****
Individuals/groups	Are the rights separable – or attached to other rights such as land?	****
Environment	Do the rights vary for classes of right holders and or with respect to time (for example, rights that may have been established under different law in time)?	*
Change/adaptation mechanisms	If so, what are the differences in the classes of rights?	*
Road to other property rights	Are rights transferable, and is there a legislative mechanism for enabling transfer?	*****
Unbundled rights	Can permanent and temporary trades take place – what is the impact of permanent trades on viability of infrastructure services along parts of the system network?	***
Risk assignment	Is trade only provided for in relation to entitlements, or can trade in derivatives take place?	**
	Can a trade be readily enforced and/or reversed if counterparty defaults?	**
	How are property rights enforced, and is the enforcement regime effective and efficient?	****
	What are the rules, if any, relating to carryover and other future period transfer of unused portion of allocations in any year?	*
	What rules/constraints attach to trading rights between connected systems?	****
	What rules attach to the technology underpinning the delivery of water to users – such as season delivery rules, channel delivery rules, etc.?	****
	Are the rights able to be qualified in any other way – and if so, on what basis?	**
	What is the risk attached to the characteristics of rights – and when does the risk materialise and can the risk be transferred with the right?	**
	How are rights presently allocated/weighed between uses – such as urban water corporations and the environment, and what interplay is there with the rights that are privately held?	*
	How do others (e.g., financial institutions/property valuers) view the value and risk profile of rights?	**
Hydrology:		
Connected systems	Is the hydrology of the system well understood, well documented, and monitored and reported on in a way that is supportive of trade?	*****
Regulated/unregulated	Is there groundwater interaction with surface water and are the interactions understood, documented, monitored and reported on?	*****
Limit and consequences of breach → environment → end of system	Are the systems modelled and is the impact of a range of future resource scenarios understood by potential market participants and regulators in relation to the system performance (both in terms of economic and environmental use)?	****
Use, including interception	Is interception of run-off included in the measurement and management of the system – or is there risk to catchments from growth in 'off stream' interception?	***
Do we know what we don't know	Have water quality and or environmental considerations the potential to cause the system to fail?	***
Salinity/water quality considerations	Is the interoperability that results from trade tested or modelled?	*****
	Big picture assessment to bring these two areas together – are the rights articulated in a way that is sympathetic to:	
	The resource constraint; and	***
	The extent to which resource knowledge is complete?	******

Issues	Questions to guide discussion/thinking	Priority
Externalities and governance considerations:	Is the administrative culture and behaviour of those involved in making decisions respected and trusted? In other words, is the governance of an area strongly impartial?	****
Institutional governance		
Sleeper/dozers	Are there rights in existence that have been inactive, that if traded into a market, may over-commit the resource?	***
Input on average vs 70% rule		
Known change of use and hydrology inputs	How does change of use impact on external environment – energy and road infrastructure, supply chains, demand for labour, etc.? And is this a pecuniary externality or a real externality (noting only the latter should be a policy concern)?	**
Unregulated 'use'		
Metering/compliance	Does the supplier have the systems, resources and technology to monitor use, and to ensure use is within licences/entitlements?	****
Reversal decisions	Can unregulated use be detected?	***
	Can water use be metered, enforced, with penalties imposed?	****
Adjustment:	Is there a sufficiently diverse (potential) market for water use in the system to facilitate trade (willing buyers/ sellers with different use profiles in terms of value add per \$ of water) and what is the likely magnitude of these gains (ex-transaction costs)?	*****
Heterogeneity → Gains from trade		
Societal pressures		
New knowledge	Is the political context mature enough to deal with trade – and accepting of the gains from trade as well as the adjustment costs in terms of activity changes that will be involved with trade?	*****
Early-mover advantage	Is there access to the skills, knowledge and finance needed to take advantage of the production possibilities afforded by access to water from trade?	*
Legislation		
Entitlement registers and accounting systems:	Has the State made plans for trade in the system, and how far advanced is the planning?	****
Legislation	Are enabling resources (e.g., registers) available/reliable/trustworthy?	*****
Plans	Is information made available on likely market conditions for trade, and is it reliable and trustworthy?	****
Registers	How mature, effective and efficient are the regulatory settings, the institutions and the services that support trades (e.g., online trading platforms)?	
Early-mover legislation		
Information availability	Have intermediaries indicated a willingness to support the function of the market?	***
Allocation announcements		
Compliance		
MER		
Intermediaries		
System type:	Which water sources in the system are capable of being made available for trade?	****
Regulated/unregulated	What is the status of infrastructure and what are the costs of accessing water in the system, and at various parts of the system?	*****
Surface water/groundwater	Does trade need to be regulated for system performance and/or economic and social interests in different parts of the system and at whose cost (benefit)?	***
Connectivity	If so – have the rules for trade been identified based on reliable data and articulated to the market and regulators?	***

3. Water markets in Africa: an analysis of Mozambique, Tanzania and Zimbabwe

Jamie Pittock, Louise Blessington, Evan W. Christen, Henning Bjornlund, Mario Chilundo, Krasposy Kujinga, Emmanuel Manzungu, Makarius Mdemu, André van Rooyen and Wilson de Sousa

3.1 INTRODUCTION

Irrigation has a potentially important role to play in alleviating poverty and food insecurity in sub-Saharan Africa (SSA). Over 40 per cent of the population of SSA lives in extreme poverty (World Bank, 2018c) and 30 per cent is food insecure (Pfister et al., 2011). With large projected population growth (World Bank, 2018a) and vulnerability to climate change (FAO, 2012), promoting irrigation is a government priority in many parts of SSA (Sullivan and Pittock, 2014). The population is predominantly rural and, despite substantial reliance on agriculture for economic growth (World Bank, 2018c), rates of agricultural productivity are low (You et al., 2010). This may be due in part to limited utilisation of irrigation (You et al., 2010); as Stirzaker and Pittock (2014) note, SSA has developed only 20 per cent of its irrigation potential. Only 4 per cent of arable land is currently irrigated (You et al., 2010), which is predicted to increase to 30 per cent by 2030 (Turral et al., 2010). Adoption of irrigation and improved irrigation techniques have many potential benefits: increased crop yields, labour savings, diversification of income sources, increased food security, improved health outcomes and greater access to education (Bjornlund et al., 2018). Irrigation accounts for 70 per cent of global water use (Wada et al., 2016), but 20–30 per cent of irrigated land is abandoned globally due to salinity and waterlogging through over-watering (Stirzaker and Pittock, 2014), so promoting water-saving initiatives is a priority.

At the same time, non-agricultural demand for water is growing, resulting in scarcity in key regions in Africa, as detailed in case studies later in this chapter. Major cities have limited water supplies, including Bulawayo (Zimbabwe), Dar es Salaam (Tanzania) and Maputo (Mozambique). The mining sector has a high demand for water, for example, in the wet–dry tropics in southern Zimbabwe. Hydropower is a vital source of electricity, but this has been curtailed in some countries due to water scarcity, for example, due to agricultural water use from the Great Ruaha River in Tanzania and droughts in Zimbabwe. Further, insufficient environmental water may reduce biodiversity, and in turn ecotourism income, such as at the Ruaha National Park in Tanzania (Pittock, 2014).

Globally, and in Africa, water scarcity is driven by climatic and other environmental changes, along with increasing demand from a growing population and different users. Effective institutions to allocate and reallocate water are needed to ensure that these limited resources are sustained, and that use generates the most benefits for society. Water markets are one means of enabling reallocation from low-value uses to those that, for example, provide more jobs and greater economic returns (Easter et al., 1999; Grafton et al., 2015). As an example, in Australia such markets enable water used to grow pasture to be sold voluntarily and with low transaction costs to farmers growing fruit or nuts: higher-value products that employ more people. To ensure equity, water markets need sound governance institutions, should be economically efficient and environmentally sustainable (Grafton et al., 2011). Yet water markets are not a panacea and are costly to implement. Many question whether the limited ability of developing countries to regulate water use, and their limited infrastructure to store and transfer water, obviates the potential for markets to succeed. Further, many people regard water as a free good and object to privatisation and commercialisation of water through markets (Meinzen-Dick, 2007).

This chapter draws on case studies from Mozambique, Tanzania and Zimbabwe, to outline the importance of irrigation in each country, current challenges and the key factors driving water use efficiency in these countries. The state of and opportunities for institutions to further evolve in each country is examined in order to assess the potential for markets to contribute to management of water scarcity. As summarised in Table 3.1, the three African countries that are the focus of this study have predominantly rural populations, who rely substantially on subsistence agriculture. Levels of poverty and food insecurity are high, but agricultural productivity and uptake of irrigation is low. Each country has set targets to expand the area of land that is irrigated (Sullivan and Pittock, 2014).

The three countries face several challenges in enhancing irrigation, including: lack of finance (Mdemu et al., 2017), poor irrigation infrastructure and

Table 3.1   Selected attributes of Mozambique, Tanzania and Zimbabwe

	Mozambique	Tanzania	Zimbabwe
% population in extreme poverty (2015 est.)[a]	69	47	21
Rural population (%, 2017 est.)[b]	67.2	67	68
Arable land (% total land, 2016)[c]	7.2	15.3	10.3
Agriculture, value added (% GDP, 2016)[d]	25	32	11
Irrigation-equipped area as % of total cultivated area[e]	2.7	3.6	5.2
% withdrawal renewable water[c]	0.7 (2015)	5.4 (2010 est.)	18 (2015)
Area of land irrigated now (ha)	27 032 (2009)[f]	363 500[g]	174 900[h]
Government target for irrigation (ha)	300 000 (by 2042)[i]	1 million[j]	2.2 million[k]

Sources: [a] World Bank (2018a); [b] World Bank (2018d); [c] World Bank (2019); [d] World Bank (2018c); [e] Svendsen et al. (2009); [f] Ministry of Agriculture (2013); [g] AQUASTAT (2016a, 2019); [h] AQUASTAT (2016b); [i] Ministry of Agriculture (2013); [j] FANRPAN (2018a); [k] FANRPAN (2018b). Adapted from Bjornlund et al. (2018).

management, poor soil fertility and lack of access to high-quality agricultural inputs (Moyo et al., 2017). Other factors, such as insecure land tenure on publicly owned schemes, and uncertainty about who should maintain infrastructure, also cause problems (Bjornlund et al., 2017), as does lack of access to extension officers who provide education and advice on agricultural and irrigation management (Markelova et al., 2009; Mdemu et al., 2017; Wheeler et al., 2017b). Secure access to markets in which to sell goods is also problematic (Mdemu et al., 2017; Moyo et al., 2017). Some government-scale attempts at irrigation and water allocation reform have proven unsuccessful (Kadigi et al., 2012), as in the case of the Great Ruaha River in Tanzania (Kashaigili et al., 2009). Competition between agriculture, hydropower and potable water supplies, combined with arguably over-ambitious plans to increase irrigation, has serious consequences. For example, cessation of Ruaha River flow in the dry season that has negatively affected wildlife-based tourism and contributed to power outages (Pittock, 2014).

More efficient use of water in irrigation may be part of the solution to managing this scarcity. However, the concept of irrigation water efficiency is contested, with strong evidence that many such programmes double-count return flows to rivers and aquifers, and deprive downstream users of access to water (Grafton et al., 2018). Proposed water efficiency programmes need to be thoroughly evaluated to understand how water access would change across scales and among different water users (Lankford, 2013). These issues are not further evaluated in this chapter, for a number of reasons. The location of the irrigation sites in these countries in the wet–dry tropics means that the base flow of rivers (if they flow at all in the dry season) is less likely to be sustained by return flows, and groundwater use is currently minimal. In most (but not all) of the examples discussed, less frequent irrigation currently results in less water being extracted. Further, given current government policies for expansion of irrigation schemes, the benefits of enhanced production from existing schemes could be debated as an alternative to expanding the area irrigated.

Within this context, we assess the role that water markets could play in helping to balance water usage between competing interests such as agriculture, urban users, the mining sector, hydropower and conservation of the environment.

3.2 CASE STUDY BACKGROUND

Drawing on the 2013–21 Australian Centre for International Agricultural Research (ACIAR) research project 'Transforming irrigation in Southern Africa' (TISA), Mozambique, Tanzania and Zimbabwe were selected as case study countries. The project focused on reforming operations of failing publicly owned smallholder irrigation schemes to be more productive and sustainable (Bjornlund et al., 2018).

Ostensibly, water availability differs in each country. Water stress per country (ratio of withdrawals to supply) by 2040 is projected to be low in Mozambique (<10 per cent), low to medium in Tanzania (10–20 per cent) and high in Zimbabwe (40–80 per cent) (Maddocks et al., 2015). However, in these countries, where most precipitation occurs in short wet seasons, the limited storage in Tanzania and Mozambique, and very limited inter-basin transfer infrastructure, results in seasonal water shortages. There is a high dependence on wet season rains that are often limited by drought. There is also little access to groundwater due to limited knowledge of available resources, unreliable electricity supplies and the high cost of independent energy systems.

We use the water market readiness assessment (WMRA) framework outlined by Wheeler et al. (2017a) to assess the extent to which each of the fundamental water market enablers are present in the three countries. The framework requires the assessment of the state of knowledge and institutional enablers with respect to: (1) property rights institutions; (2) hydrology; (3) governance of externalities; (4) the types of systems; (5) water adjustment mechanisms; and (6) the state of entitlement registers and water accounting.

3.3 RESULTS

Our assessment of the extent to which each of the fundamental water market enablers are present in Mozambique, Tanzania and Zimbabwe is summarised in Table 15.1, in Chapter 15. Elaborating on these results, trends and examples are outlined for each of the fundamental market assessors. Each of the countries has strong property rights institutions with national water laws that separate water entitlements from land ownership. The rights are not directly transferable among users, requiring unused water entitlements to be returned to a water agency for reallocation. While water agencies in each country have a legal mandate to enforce water allocations, their actual capacity to do so is curtailed by limited resources or by local politics. We have observed active enforcement of rules for access to water from public infrastructure by the Zimbabwe National Water Authority (ZINWA). In Tanzania, we have observed the opposite: with farmers installing their own pumps and irrigating directly from the river, the river basin boards appearing to lack the resources and authority to control agricultural diversions from rivers. In all three countries, urban and 'basic' domestic use is prioritised ahead of environmental and agricultural use. Recently, in southern Mozambique, irrigated agricultural use was suspended in order to supply water to Maputo.

Understanding of the hydrology of the water systems is patchy. There is hydrological information available and there are models for surface water systems, but groundwater systems are very poorly assessed. As water scarcity is common in all three countries, it is understood that trade-offs will need to be made in allocating water between different users. Zimbabwe does have a National Water Masterplan and other processes for managing surface water, including outline plans and caps on extraction for each system. In Tanzania, when issuing water rights, the river basin boards consider: (1) already committed water; (2) availability of water; and (3) water needed for environmental use. Water use permits can be restricted on account of insufficient water, droughts and natural disasters. Adaptation to the impacts of climate change on water availability are yet to be incorporated into governance in the three countries.

Regarding the types of systems, there is relatively little infrastructure that allows transfer of water allocations to other users and river basins. Most reservoirs in Mozambique and Tanzania are for hydropower and urban water supply, with agricultural users extracting water from rivers. Zimbabwe has a colonial legacy of reservoirs supplying water for irrigation and one inter-basin transfer canal. The cost of building non-hydropower water storage appears prohibitive for these governments. As there is little infrastructure to enable trade, relevant regulations have not been developed.

Governance of externalities is patchy. While there are sound regulations, their application is often hindered by lack of impartiality or limited resources. Cross-sectoral harmonisation of policy among government water, agriculture, energy and other sectoral agencies is often challenging. For example, in Tanzania, the river basin boards are poorly resourced and have different administrative boundaries compared with the powerful district governments that prioritise economic development, especially in agriculture.

Currently, in the event of scarcity the water adjustment mechanisms involve administrative intervention, because there are no business-to-business trading mechanisms. This results, for example, in suspending irrigation diversions without compensation in order to provide water for Maputo in drought. Informal discussion is under way in Zimbabwe among water experts and government officials who recognise a range of policy options for enhancing water allocations. Entitlement registers and capacity for water accounting are being established in the three countries. Zimbabwe has a water entitlement register, but it is not publicly accessible.

While there are some key building blocks in place (trade step 1 of 3 in Wheeler et al.'s, 2017a WMRA framework), we find that at present Mozambique, Tanzania and Zimbabwe do not have the institutional and governance structures that would be necessary for effective formal water markets. If formal water markets are not currently feasible, the question is whether alternative administrative, voluntary or informal markets may be sufficient to help these societies manage water scarcity in the immediate future. Administrative reallocation of water and enforcement of efficiency rules are blunt instruments that are constrained by the limited capacities of the government agencies and the interests of current water users.

We propose instead that various push and pull factors are the key drivers of water use in the countries studied. Bjornlund et al. (2018) show how the provision of soil monitoring tools and farmer learning has resulted in a voluntary reduction in over-irrigation and nutrient leaching. For example, the interval between irrigation events increased by between 1.4 and 9 days per scheme (Bjornlund et al., 2018). The reduced labour required and increased crop yields from this approach appear to have been the main drivers of water use (Bjornlund et al., 2018), rather than high-level water trading schemes, especially as individual farmers do not pay volumetric fees for their water use. The question is whether farmers in these developing countries could be further rewarded for wider uptake of these better agronomic and more water-efficient practices through development of informal water markets or case-by-case programmes. Examples from each of the three countries are now explored.

3.3.1 Catalysts for Trade?

While formal water markets will not be established in these countries in the foreseeable future, there are local opportunities to facilitate other informal transactions among farmers and between irrigation schemes for urban, industrial and environmental uses. There is great potential for transactions among individual farmers, such as in the pumped schemes in Mozambique. As an example, in Mexico, farmers within a scheme at times of scarcity mutually agree who is not going to farm that season and how much those who are going to farm are going to pay those who are not (Bjornlund and McKay, 2002). In each of the three countries there are areas where there is water scarcity and demand from sectors that place a high socio-economic value on access to more water. These situations might be catalysts for establishing innovative water trades, as the following case studies indicate.

3.3.2   Zimbabwe

The Silalatshani (also called the Silabuhwa) Dam in the Mzingwane River catchment (a tributary of the Limpopo River) provides water for irrigation, gold mines and urban users in the west Nicolson area of Zimbabwe (FAO, 2017). Currently ZINWA supplies irrigation schemes with 'agreement water' (a bulk water supply contract) from the dam on an annual basis.

The TISA research project deployed simple-to-use soil moisture and fertility monitoring tools that have seen farmers change their behaviour and irrigate around half as often as they did before. In the Mkoba and Silalatshani (Landella block) irrigation schemes, grain and maize water productivity increased, respectively, from 0.19 kg/m^3 and 0.20 kg/m^3 in the 2013/14 season to 0.69 kg/m^3 and 0.95 kg/m^3 for farmers without the monitoring tools, and to 1.00 kg/m^3 and 1.28 kg/m^3 for farmers with the tools (Moyo et al., 2020).

ZINWA supplies and charges water to the scheme rather than to individual farmers. Consequently, the incentive for farmers to use less water is not to lower water costs, but to achieve more effective use of fertilisers due to reduced leaching, greater crop yields and reduced labour that can be focused on other livelihood activities. Before the project interventions with soil monitoring tools and institutions to improve profitability from the value chain, the farmers regarded the water fees as too expensive and did not pay them, contributing to system failure. Since project interventions have made irrigation profitable, the farmers have paid the water fees and the irrigation system is functioning. However, the cost of the soil monitoring tools (that are not yet in commercial production) is currently more than the farmers say that they are willing to pay (Abebe et al., 2020).

This raises the question as to how the water that farmers are paying for but not using should be allocated. Options include expanding the irrigation area or reallocating the water to the environment or other consumptive users, such as the mining industry. Under current Zimbabwean law, farmers cannot sell their excess entitlement to another water user, but could choose to surrender or lease part of their agreement water to ZINWA and pay proportionally lower water fees. Farmers would have a larger incentive to use water more efficiently if they received more direct and ongoing benefits. For example, miners might wish to provide incentives for farmers to save water. ZINWA would sell the extra water to the miners, and the farmers could benefit. The incentives could include payments or investments in more efficient irrigation infrastructure and soil monitoring tools. The broader impacts on local, provincial and national economies and local communities would need to be investigated to fully understand the trade-offs.

3.3.3   Mozambique

In Mozambique the Strategic Plan for Agricultural Development (2011–20) clearly indicates that with regard to irrigation policy and programme there is (or should be) close collaboration between the Ministry of Public Works and Housing and the Ministry of Agriculture (now Ministry of Agriculture and Food Security) around the use of water resources for agriculture (Ministry of Agriculture, 2011). This is in order to manage competition for water resources across sectors. Competition for water from the Pequenos Libombos Dam (PLD), which supplies water for irrigation and Maputo city, is an example of this competition. The PLD, constructed in 1987, is located about 45 km southwest of Maputo city on the Umbeluzi River. The PLD has a storage capacity of 400 Mm³ and is the major water supplier to Maputo metropolitan area feeding the water treatment plant on the Umbeluzi River. Although the dam was constructed mainly to supply water to Maputo city and surrounding areas, it was also expected to support 3500 hectares of irrigation and generate 1.72 MW of hydropower. Secondary benefits have been fishing and tourism. However, the Umbeluzi River is part of the Inkomati basin, which straddles Mozambique, South Africa and Swaziland. Mozambique being the downstream country has historically struggled to get adequate water. A review of the situation in 2008 stated that water allocation in Umbeluzi River faced considerable difficulties as it was predicted that future demands could not be fully met even with an increase in reservoir storage (Juízo and Lidén, 2008). This situation has been exacerbated by climate change and drought. Detailed analysis of the Intergovernmental Panel on Climate Change's (IPCC) global circulation models predict drier conditions for northeastern South Africa and more extreme events, and the food security cluster analysis predicts a decline in rainfall of 10–15 per cent in the Maputo region in the next 30 years (Engelbrecht et al., 2009).

There are two distinct seasons, dry (March–October) and wet (November–February). Therefore, the situation alternates between heavy rain and flooding, and completely dry. The ongoing difficulty then lies in managing these extremes. However, a major drought in the region has meant that inflows are dwindling. In February 2019 there remained only 12 months water supply for

Maputo city in the PLD, and only 80 per cent of consumer demand and 15 per cent of farmer demand was being met (Marinela, 2019).

Farmers involved in the TISA project in the '25 de Setembro' irrigation scheme on the Umbeluzi River downstream of the PLD have been able to significantly reduce their water application, whilst at the same time increasing their yields. Chilundo et al. (2020) show that, for example, green maize smallholder farmers (1–2 ha) have reduced their number of irrigations by 30 per cent, whilst increasing yield by up to 300 per cent. This has been due to improved irrigation infrastructure, irrigation water management and agronomy. The reduction in water used per crop has been partly offset as abandoned fields have been brought back into production and irrigation has become more profitable. The TISA project has assessed water use and gross margins for green maize under the 25 de Setembro scheme. The gross margin per m³ of irrigation water for green maize (one of the highest-value crops) was found to range from 0.4 to 0.9 US cents/m³ before the project intervention. This was due to very low yields, and high levels of water use. After the project intervention the best farmers get as much as 39–56 US cents/m³; however, typical values were less than half that.

In comparison, the retail price of potable water, delivered by Àguas de Regiao de Maputo (ARM), in 2018 was around 69 US cents/m³. Clearly this is not the gross margin value but it does show the large difference compared to farmers' previous gross margin of just 0.4–0.9 US cents/m³, which post-intervention is closer to the current potable retail price. Therefore, rather than cutting off all irrigation water to farmers in periods of water scarcity, a more equitable approach may involve ARM supporting the farmers wanting to irrigate. ARM could provide the farmers with the support that, for example, the TISA project provided, to use less water more efficiently, or perhaps even pay the farmers for not taking water. This would start to develop a greater appreciation of the value of water and spread the impacts of water scarcity more equitably between the agricultural and urban sectors.

3.3.4   Tanzania

Irrigation diversions in the upper Ruaha River basin have resulted in the river ceasing to flow in the dry season through the Ruaha National Park, negatively impacting upon wildlife and tourism (Stommel et al., 2016) and greatly reducing hydropower generation from the Mtera and Kidatu hydropower dams that supply half Tanzania's electricity (Pittock, 2014). Power cuts have negatively impacted upon the Tanzanian economy. For many years Tanzanian political leaders have promised to restore river flows, without success (Kashaigili et al., 2009).

As with Mozambique and Zimbabwe, interventions within the irrigation schemes in the Ruaha River basin have greatly reduced water application and increased crop productivity. Under the Kiwere scheme, those with soil monitoring tools reduced their time spent irrigating by 65 per cent, while their neighbours reduced it by 47 per cent, and green maize yields increased by an average of 28 per cent (Mdemu et al., 2020). However, the project has not looked beyond the scheme scale, where there are likely to be broader benefits from improving smallholder irrigators' productivity while reducing irrigation water use. Hydro-economic modelling could be used to explore various investments, including trade-offs between investing in more productive irrigation and provision of more water for the tourism industry in the national park and hydropower production.

3.4   DISCUSSION AND CONCLUSION

These cases raise the question of whether investments in more productive water use in irrigation genuinely free up water for use in other sectors, as opposed to reducing return flows to rivers and enabling expansion of agricultural production (Grafton et al., 2018). In the cases cited above there have been countervailing outcomes, such as previously disused irrigation plots being brought back into production on some schemes. Each water efficiency investment does need to be thoroughly assessed to identify and manage double-counting of water and potentially perverse outcomes. However, in many of the TISA irrigation schemes in the wet-dry tropics, before intervention there was no obvious beneficiary from return flows, as it went to rivers with no surface flows in the dry season in Zimbabwe, nor to groundwater users.

Trading within irrigation schemes may also be beneficial. In the Magozi rice scheme in Tanzania, conflicts over water use have been at a high level, especially between upstream and downstream users, with some downstream users unable to grow a crop. With the reduction in conflicts experienced during the TISA project, there might be sufficient willingness to engage in collective action during future water shortages, so collective agreements could be made at the Irrigation Association level about which farmers will not sow crops in a given season and how they will be compensated. Such practices are used in some schemes, for example, in the Delicias system in Mexico (Bjornlund and McKay, 2002).

Payment of water fees has proven to be a critical issue in many irrigation schemes in Africa, with non-payment or non-participation in maintenance work resulting in dilapidated and under-utilised infrastructure. Oman provides an example of how a special version of water markets could help to overcome this problem. In some Omani systems the irrigation organisation controls a proportion of the water right. The proportion is auctioned to farmers on a weekly, monthly or annual basis and the proceeds pay for operations and maintenance (Zekri et al., 2006). Such systems do not require a fully operational formal market system, as discussed in this book.

Experiences from other parts of the world suggest that gaining experience of water trading through informal or pilot market arrangements is likely to increase confidence in introducing more formal mechanisms. It should also be noted that in some places where formal markets have been introduced they have not been very active. However, these markets may facilitate individual critical transactions that might have significant positive economic impacts, such as by allowing new urban developments or the construction of processing industries for the farming industry. This has been the case in Chile (Bjornlund and McKay, 2002) and Alberta, Canada (Bjornlund and Klein, 2015).

While the situation is likely to change over time, we conclude that formal water markets are not currently suited to managing water scarcity in Mozambique, Tanzania and Zimbabwe, as institutional and governance structures are insufficient, and challenges such as lack of available inputs, knowledge and market access persist. Each country has good knowledge of surface water resources, and sound laws that separate water access from land titles. However, currently no country allows trade of water entitlements, and knowledge of the use and management of groundwater is poor. Government agencies lack the resources to manage water as transparently and effectively as they would wish. In Zimbabwe, great water scarcity has led to discussion among government officials and water experts of many of the concepts and opportunities that may arise from water markets.

Experience from the TISA project suggests that, at present, to provide incentives for farmers to irrigate more efficiently we should instead look to interventions that save irrigation farmers time and energy costs, and maximise crop yield so that, in some cases, total water use is reduced (Bjornlund et al., 2018). This opens opportunities for informal water market arrangements to emerge between farmers and between irrigation organisations and external parties such as mines, hydropower, national parks (tourism) and other industries. In this chapter we have provided examples of such potential opportunities identified during the TISA project. Demands from other sectors may well push forth future, more formal collaborations among stakeholders to equitably reallocate water to uses that have higher socio-economic returns, as illustrated by the following examples. In Mozambique, water shortages in the capital Maputo may strengthen calls for agricultural water efficiency. In Tanzania, there is already tension between various users of water from the Great Ruaha River, which is in demand for agriculture, tourism and hydropower. As Zimbabwe's water-intensive mining sector grows, this may increase competition for water resources from its main consumer in the agricultural sector. This competition for water in sectors across the three countries may catalyse innovation and a transition to more structured methods for reallocating water, including the potential development of water markets.

REFERENCES

Abebe, F., Wheeler, S.A., Zuo, A., Bjornlund, H., Van Rooyen, A., et al. (2020). Irrigators' willingness to pay for the access to soil moisture monitoring tools in South Eastern Africa. *International Journal of Water Resources Development*, 36(Suppl. 1), 246–267.

AQUASTAT (2016a). United Republic of Tanzania. http://www.fao.org/nr/water/aquastat/countries_regions/TZA/index.stm.

AQUASTAT (2016b). Zimbabwe. http://www.fao.org/nr/water/aquastat/countries_regions/ZWE/.

AQUASTAT (2019). United Republic of Tanzania irrigated area sheet. http://www.fao.org/nr/water/aquastat/countries_regions/index.stm.

Bjornlund, H., and Klein, K.K. (2015). Water conservation and trading – policy challenges in Alberta Canada. In K. Burnett, R. Howitt and J. Roumasset (eds), *Handbook of Water Economics and Institutions* (pp. 381–396). Routledge.

Bjornlund, H., and McKay, J. (2002). Aspects of water markets for developing countries: experiences from Australia, Chile, and the US. *Environment and Development Economics*, 7(4), 769–795.

Bjornlund, H., Parry, K., Pittock, J., Stirzaker, R., van Rooyen, A., et al. (2018). Transforming smallholder irrigation into profitable and self-sustaining systems in southern Africa. In *Smart Water Management* (pp. 330–387). Korean Water Resources Corporation and International Water Resources Association.

Bjornlund, H., van Rooyen, A., and Stirzaker, R. (2017). Profitability and productivity barriers and opportunities in small-scale irrigation schemes. *International Journal of Water Resources Development*, 33(5), 690–704.

Chilundo, M., de Sousa, W., Christen, E., Faduco, J., Bjornlund, H., Cheveia, E., Munguambe, P., Jorge, F., Stirzaker, R. and Van Rooyen, A. (2020). Do agricultural innovation platforms and soil moisture and nutrient monitoring tools improve the production and livelihood of smallholder irrigators in Mozambique? *International Journal of Water Resources Development*, 36(S1), S127–S147.

Easter, K.W., Rosegrant, M.W., and Dinar, A. (1999). Formal and informal markets for water: institutions, performance, and constraints. *World Bank Research Observer*, 14(1), 99–116.

Engelbrecht, F., McGregor, J., and Engelbrecht, C. (2009). Dynamics of the Conformal-Cubic Atmospheric Model projected climate-change signal over southern Africa. *International Journal of Climatology: A Journal of the Royal Meteorological Society*, 29(7), 1013–1033.

FANRPAN (2018a). *Pathways for Irrigation Development: Policies and Irrigation Performance in Tanzania*. https://www.fanrpan.org/sites/default/files/publications/Hi_Res_Tanzania14_02_17.pdf.

FANRPAN (2018b). *Pathways for Irrigation Development: Policies and Irrigation Performance in Zimbabwe*. https://www.fanrpan.org/sites/default/files/publications/Hi_Res_Zimbabwe14_02_17.pdfWater.

FAO (2012). *Coping with Water Scarcity: An Action Framework for Agriculture and Food Security*. Water Report. Food and Agriculture Organization of the United Nations.

FAO (2017). *AQUASTAT.* Food and Agriculture Organization of the United Nations, Rome. Accessed 5 April 2021 at http://www.fao.org/nr/water/aquastat/main/index .stm.

Grafton, R.Q., Horne, J., and Wheeler, S.A. (2015). On the marketisation of water: evidence from the Murray–Darling Basin, Australia. *Water Resources Management, 30*(3), 913–926.

Grafton, R.Q., Libecap, G., McGlennon, S., Landry, C., and O'Brien, B. (2011). An integrated assessment of water markets: a cross-country comparison. *Review of Environmental Economics and Policy, 5*(2), 219–239.

Grafton, R.Q., Williams, J., Perry, C.J., Molle, F., Ringler, C., et al. (2018). The paradox of irrigation efficiency. *Science, 361*(6404), 748–750.

Juízo, D., and Lidén, R. (2008). Modeling for transboundary water resources planning and allocation. *Hydrology and Earth System Sciences Discussions, 5*(1), 475–509.

Kadigi, R.M.J., Tesfay, G., Bizoza, A., and Zanabou, G. (2012). Irrigation and water use efficiency in Sub-Saharan Africa. Policy Research Paper 4. New Delhi, India, GDN.

Kashaigili, J.J., Rajabu K., and Masolwa, P. (2009). Freshwater management and climate change adaptation: experiences from the Great Ruaha River catchment in Tanzania. *Climate and Development, 1*(3), 220–228.

Lankford, B. (2013). *Resource Efficiency Complexity and the Commons: The Paracommons and Paradoxes of Natural Resource Losses, Wastes and Wastages.* Routledge.

Maddocks, A., Young, R.S., and Reig, P. (2015). Ranking the world's most water-stressed countries in 2040. World Resources Institute, http://www.wri.org/ blog/2015/08/ranking-world%.

Marinela, C. (2019). Enough water for 12 months only as Pequenos Libombos faces worst drought in 30 years. https://clubofmozambique.com/news/enough-water-for -12-months-only-as-pequenos-libombos-faces-worst-drought-in-30-years-watch/.

Markelova, H., Meinzen-Dick, R., Hellin, J., and Dohrn, S. (2009). Collective action for smallholder market access. *Food Policy, 34*(1), 1–7.

Mdemu, M., Kissoly, L., Bjornlund, H., Kimaro, E., Christen, E. W., van Rooyen, A., Stirzaker, R. and Ramshaw, P. (2020). The role of soil water monitoring tools and agricultural innovation platforms in improving food security and income of farmers in smallholder irrigation schemes in Tanzania. *International Journal of Water Resources Development, 36*(S1), S148–S170.

Mdemu, M.V., Mziray, N., Bjornlund, H., and Kashaigili, J.J. (2017). Barriers to and opportunities for improving productivity and profitability of the Kiwere and Magozi irrigation schemes in Tanzania. *International Journal of Water Resources Development, 33*(5), 725–739.

Meinzen-Dick, R. (2007). Beyond panaceas in water institutions. *Proceedings of the National Academy of Sciences, 104*(39), 15200–15205.

Ministry of Agriculture (2011). *Plano estrategico para desenvolvemento do sector agrario.* Maputo, Mozambique.

Ministry of Agriculture (2013). *Estratégia de Irrigação.* Maputo, Mozambique.

Moyo, M., Van Rooyen, A., Bjornlund, H., Parry, K., Stirzaker, R., et al. (2020). The dynamics between irrigation frequency and soil nutrient management: transitioning smallholder irrigation towards more profitable and sustainable systems in Zimbabwe. *International Journal of Water Resources Development, 36*(Sup. 1), S102–S126. DOI: 10.1080/07900627.2020.1739513.

Moyo, M., van Rooyen, A., Moyo, M., Chivenge, P., and Bjornlund, H. (2017). Irrigation development in Zimbabwe: understanding productivity barriers and opportunities at Mkoba and Silalatshani irrigation schemes. *International Journal of Water Resources Development*, 33(5), 740–754.

Pfister, S., Bayer, P., Koehler, A., and Hellweg, S. (2011). Projected water consumption in future global agriculture: scenarios and related impacts. *Science of the Total Environment*, 409(20), 4206–4216.

Pittock, J. (2014). Why water and agriculture in southern Africa? In J. Pittock, R. Grafton and C. White (eds), *Water, Food and Agricultural Sustainability in Southern Africa* (pp. 1–8). Tilde Publishing & Distribution.

Stirzaker, R., and Pittock, J. (2014). The case for a new irrigation research agenda for Sub-Saharan Africa. In J. Pittock, R.Q. Grafton and C. White (eds), *Water, Food and Agricultural Sustainability in Southern Africa* (pp. 91–107). Tilde University Press.

Stommel, C., Hofer, H., and East, M.L. (2016). The effect of reduced water availability in the Great Ruaha River on the vulnerable common hippopotamus in the Ruaha National Park, Tanzania. *PLoS ONE*, 11(6), e0157145. DOI: 10.1371/journal.pone.0157145.

Sullivan, A., and Pittock, J. (2014). Agricultural policies and irrigation in Africa. In J. Pittock, R.Q. Grafton and C. White (eds), *Water, Food and Agricultural Sustainability in Southern Africa* (pp. 30–54). Tilde University Press.

Svendsen, M., Ewing, M., and Msangi, S. (2009). Measuring irrigation performance in Africa. International Food Policy Research Institute, Washington, DC.

Turral, H., Svendsen, M., and Faures, J. (2010). Investing in irrigation: reviewing the past and looking to the future. *Agricultural Water Management*, 97, 551–560.

Wada, Y., Flörke, M., Hanasaki, N., Eisner, S., Fischer, G., et al. (2016). Modeling global water use for the 21st century: the Water Futures and Solutions (WFaS) initiative and its approaches. *Geoscientifc Model Development*, 9(1), 175–222.

Wheeler, S.A., Loch, A., Crase, L., Young, M., and Grafton, R.Q. (2017a). Developing a water market readiness assessment framework. *Journal of Hydrology*, 552, 807–820.

Wheeler, S.A., Zuo, A., Bjornlund, H., Mdemu, M., van Rooyen, A., and Munguambe, P. (2017b). An overview of extension use in irrigated agriculture and case studies in south-eastern Africa. *International Journal of Water Resources Development*, 33, 755–769.

World Bank (2018a). Poverty. https://data.worldbank.org/topic/poverty?locations=ZG.

World Bank (2018b). Population projections and estimates. http://databank.worldbank.org/data/reports.aspx?source=health-nutrition-and-population-statistics:-population-estimates-and-projections.

World Bank (2018c). Agriculture, forestry, and fishing, value added (% of GDP). https://data.worldbank.org/indicator/NV.AGR.TOTL.ZS?locations=ZG.

World Bank (2018d). Rural population (% of total population). https://data.worldbank.org/indicator/SP.RUR.TOTL.ZS.

World Bank (2019). Arable land (% of land area) – Mozambique, Zimbabwe, Tanzania. https://data.worldbank.org/indicator/AG.LND.ARBL.ZS?locations=MZ-ZW-TZ.

You, L., Ringler, C., Nelson, G., Wood-Sichra, U., Robertson, R., et al. (2010). *What is the Irrigation Potential for Africa? A Combined Biophysical and Socioeconomic Approach.* International Food Policy Research Institute.

Zekri, S., Kotagama, H., and Boughanmi, H. (2006). Temporary water markets in Oman. *Journal of Agricultural and Marine Sciences*, 11, 77–84.

4. Agricultural water markets in China: a case study of Zhangye City in Gansu province

Alec Zuo, Tianhe Sun, Jinxia Wang and Qiuqiong Huang

4.1 INTRODUCTION

Irrigation agriculture in China is of critical importance to national food security. Over 70 per cent of China's staple crops, around 80 per cent of cash crops, and 90 per cent of vegetables are produced by irrigated agriculture (Wang et al., 2017). The rapid expansion of irrigation during the twentieth century has also seen a serious degradation of the ecological environment (Sun et al., 2016). Compounding the sharp increase in demand for water is the limited and continuous reduction in water supply, which has led to a shortage of agricultural water resources, particularly in northern China (Wang et al., 2017). In response, the Chinese government at various levels has implemented a number of policy initiatives to improve water efficiency, reduce overall agricultural water use, and reallocate water between agricultural areas (Wang et al., 2019b). The Chinese government has joined a global trend towards the adoption of market-based mechanisms in order to address water scarcity (Moore, 2014).

The simplest such mechanism is water reform that increases the cost of water to encourage conservation and a transition towards higher value-added water uses (Moore, 2014). However, price increases to date have been very gradual and largely confined to urban areas, while the charge for water use in the agricultural sector – which accounts for around 65 per cent of total water withdrawals – remains very low (Zhong and Mol, 2010). Recent research has cast doubt on whether moderate increases in charges alone can significantly reduce water use in the Chinese agricultural sector, given price elasticity of agricultural water demand (Aregay et al., 2013) and negative impacts of increasing charges on farmers' income (Huang et al., 2010; Wang et al., 2016). Given the limitations of water reforms on water charges, the Chinese government has highlighted its intention to utilise another market-based mechanism, namely developing water markets (Xia and Pahl-Wostl, 2012, p. 448). Recently, the Ministry of Water Resources (MWR, 2016) issued regulations around the transferring of water rights, with the intention to establish and improve a water rights system; promote water rights trading; cultivate water markets; encourage various forms of trading; and promote water conservation, security and efficient allocation.

Water trading enables water users to buy or sell their usage to other users, and an effective water trading system is preceded by a cap on total water use within a given region, resulting in an efficient allocation of water resources whereby scarce resources are utilised only by those who value them most (Howe et al., 1986; Chong and Sunding, 2006; Debaere et al., 2014). There has been substantial progress in the development of Chinese water markets during recent years. The majority of water trading in China, however, has occurred between regions and between sectoral uses (Wang et al., 2019b), while water trading among farmers has been limited. Inter-sectoral trades are usually between agriculture (the seller) and industry or urban water utilities (the buyer). In the long term, the development of formal trading among irrigation water users will be pivotal in achieving fully functional agricultural water markets. However, the suitability of water markets within the context of a developing country such as China has been contested. While the development of water markets aspires to achieve an economically efficient allocation of water resources, the distribution of these resources is not necessarily equitable, which in turn raises concerns about their impact on disadvantaged water users (Johansson et al., 2002; Araral, 2010; Ray, 2013). Other concerns include the effect of consumptive use on downstream users, the need to ensure minimum flows for environmental purposes (Gould, 1988; Lui, 2004; Etchells et al., 2006), as well as the risk of large-scale water rights transfers from rural to urban areas, potentially destabilising rural communities (Chong and Sunding, 2006; Grafton et al., 2012). Therefore, in order to judge whether China's agricultural water sector is ready for water trading or what conditions need to be met before implementation, a detailed evaluation of current water trading cases within China is necessary. The aim of this chapter is to harness the water market readiness framework (Wheeler et al., 2017) to assess a water market case study in China.

4.2   EXISTING LITERATURE AND BACKGROUND: WATER MARKETS IN CHINA

A number of studies have highlighted the challenges and pitfalls of introducing water markets in China, given that water rights are neither clearly defined nor effectively enforced (e.g., Calow et al., 2009; Speed, 2009; Sun, 2009).

Recently the basic institutional conditions, such as the assignment of water use rights to individual users for market-based allocation of water resources, have been dramatically strengthened. Several pilot and model projects have been established, such as the first formal water trade between two counties (namely, Dongyang and Yiwu in Zhejiang province), and spot transfers of water based on water certificates between farmers in Zhangye (Jiang, 2018). Perhaps most importantly, the Ministry of Water Resources of China is committed to a major expansion of water rights trading within seven provinces across China (Moore, 2014). Despite this, trading in China remains relatively confined to small areas, rather than entire provinces, and largely entails inter-jurisdictional/ sectoral rights transfers, as opposed to a market exchange of water quantity trading between irrigated water users, or the common form of markets advocated by economists (Gao, 2007; Lewis and Zheng, 2018).

Moore (2014) presents the initial challenge to effective water markets in China as one of definition, knowledge and communication. Chinese water resource specialists have frequently stressed the unclear legal foundations of water trading in China, including a failure (in many areas) to clearly define annual water use entitlements, along with the duration of granted water rights. This is a particular issue when pilot water trading projects end after a certain period of time, leaving water right holders unclear whether they must reapply to continue their water allocations (Gao, 2007). Integrated coordination and management between the administrative jurisdictions and bureaucratic units responsible for water resource management is also a fundamental barrier in encouraging the expansion of water markets in China (Moore, 2014). For example, in some regions it is unclear which entity is responsible for water trading across provinces; whereas separate bureaucratic units are responsible for regulating surface and groundwater, undermining efforts to integrate these resources into water trading systems (Gao, 2007).

Another major challenge to water markets in China is the monitoring and enforcement of water use, with compliance particularly weak at the individual water user level. As a result, over-consumption of water is rife (Moore, 2014). According to Chinese water resource experts, abstraction permits for wells are rarely obtained and, given the sheer number of wells across China, enforcement is virtually non-existent (Wang et al., 2019a). Furthermore, those users who do obtain abstraction permits regularly exceed their quotas, with often no effective means of ensuring these quotas are followed. This is particularly the case in northern China, where groundwater withdrawal accounts for a high percentage of total water extraction (Gao, 2007). With regards to metering, Lewis and Zheng (2018) suggest that the most practical steps initially would be to work with the current traditional estimation methods; however, in a much stricter form, for example by taking more explicit measurements of water depths in canals, and by keeping more thorough and open records. Ultimately, in the same manner as Water User Associations (WUAs),[1] increased transparency should lead to fewer arguments, and more acceptance of how water is distributed (Lewis and Zheng, 2018).

Lewis and Zheng (2018) consider the challenges behind obtaining stronger water rights and more accurate water metering, highlighting the fact that the available water is being shared among a very high number of farmers, each getting relatively tiny amounts of water for generally small plots of land. However, notwithstanding the tiny amounts of water per farm and the tight margins involved, water is still being applied unnecessarily, given the comparatively low water use efficiency across China (Zhu et al., 2001; Xie, 2009). As highlighted by the experience in Australia during extreme water shortages, a more effective market might well help in getting the small amounts of water that are available to their best use (Lewis and Zheng, 2018). In 2016, the Chinese Ministry of Water Resources and the Beijing Municipal Government (with the approval of the State Council) established the China Water Exchange, in order to promote water trading across China. Since 2016, a number of transactions have occurred through the exchange (Table 4.1). The exchange lists two types of water trades: inter-sectoral trades and those within the irrigation sector. While most trades occur within one province, there are a number of transactions between provinces and the capital city, Beijing. Transfer terms for inter-sectoral trades vary from 1 to 25 years and are usually determined through negotiations between the seller and buyer. The price of water is also determined by negotiation, and long-term transfers are usually more expensive than shorter-term transfers. Volumes traded among irrigation users are at a much smaller scale than inter-sectoral trades. Some trades occur between WUAs, while other transactions are between individual irrigators. For trades within the irrigation sector, all transfer terms are for one year, and average prices range between 60 and 385 RMB/ML.[2]

4.3 WATER MARKET READINESS ASSESSMENT: A CASE STUDY

According to the water market readiness assessment (WMRA) framework developed by Wheeler et al. (2017), there are three steps of evaluation. The first of these is a background check, involving an evaluation of current institutional, legislative, planning and regulatory capacity to facilitate and/or enable water trades. The second step is to evaluate whether the development and implementation of a water market is effective and efficient, including aspects such as market governance, compliance and enforcement. The final step is monitoring and an ongoing review/assessment, mainly involving further development of the trade-enabling mechanisms, such as reducing transaction

Table 4.1   *Annual summary of water trading listed on the China Water Exchange*

Province	Year	Total volume traded (ML)	Current price (Chinese yuan/ML)	Term (year)
Inter-sectoral water trades				
Guizhou	2019	491.8	280	1
Inner Mongolia	2018	1 276 165	600 (first payment)	25
Beijing (seller from Shanxi)	2018	41 900	350	1
Inner Mongolia	2017	457 800	600 (first payment)	25
Beijing (seller from Hebei)	2017	40 000	218	1
Henan	2017	300 000	760	3
Inner Mongolia	2016	500 000	600 (first payment)	25
Shanxi	2016	9 000	1200	5
Henan	2016	24 000	870	3
Beijing (seller from Hebei)	2016	70 410	278	1
Ningxia	2016	32 853	931	15
Water trades within the irrigation sector				
Hunan	2019	4 298.2	90.21	1
Shanxi	2019	24.2	384.69	1
Gansu	2019	260.0	60.00	1
Hebei	2019	131.2	200.00	1
Shandong	2018	0.6	175.00	1
Ningxia	2018	400.0	100.00	1
Xinjiang	2017	6 253.1	216.00	1
Ningxia	2017	2 600.0	227.23	1
Hebei	2017	310.9	200.00	1

Source: Data from China Water Exchange (http://cwex.org.cn/lising), water trades are collected as of 31 July 2019.

costs, adapting to new information, accounting for unanticipated externalities and developing refined market products.

4.3.1   Case Study Area

Zhangye City (see Figure 4.1), located in the middle reaches of the Heihe River Basin (HRB), is one of the top ten commodity grain production areas in China. The HRB is the second-largest inland river basin in China, located in the arid area of northwestern China. It covers an area of approximately 116 000km^2 with 0.26 million ha of arable land and has a mean runoff of 2800

Mm³/a. The upper, middle and lower reaches of the basin have very different geomorphological characteristics. The upper reaches lie in mountainous areas, the middle reaches have rich soils and flat plains, and the lower reaches are mostly desert. In 2014, total water withdrawal was 3.19 billion m³, with irrigated agriculture accounting for 65.3 per cent of total withdrawal (2.08 billion m³). With the rapid socio-economic development in the middle reaches of the river basin, water shortage in the basin has become more and more acute.

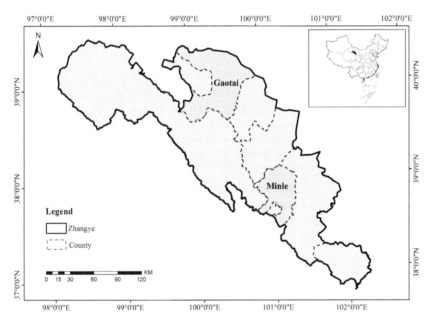

Figure 4.1   Location of Zhangye City, Gaotai and Minle counties

This study uses a semi-structured in-depth interview approach to collect qualitative information from key stakeholders in the case study area. In August 2016, two researchers conducted ten semi-structured in-depth interviews with officials from the Heihe River Basin Authority, Zhangye City Government, Mingle County Government and Hongshuihe Irrigation District. The interviews were undertaken in a top-down manner from the basin level to subsequent city, county and irrigation district level, so that comprehensive information on water market conditions and development could be collected. The interview checklist was based on the water market assessment questions from Appendix A in Wheeler et al. (2017).

4.3.2   Water Market Development in Zhangye City

In Zhangye, although water rights have been granted to individual farmers in the form of water right certificates since 2002, and informal water trading among farmers has been around for decades, the government officially announced water right transaction policies under comprehensive reforms of agricultural water rights and charges. In 2015, the Zhangye Municipal People's Government requested its counties to renew water right certificates, promote water right transfers and start to establish water right markets. The official administrative measures on water rights trading were introduced in 2016, which specified the scope, content, procedures and pricing rules of water rights trading. For transactions among irrigation users, it included trading between individual farmers or WUAs. In addition, county water authorities could establish a water rights repurchase system to encourage water saving. In 2017, more detailed management measures for water rights trading were issued in each of Zhangye's counties.

Presently there is no active water market among water use sectors or within the agricultural sector. However, water transfers do occur informally or formally under signed contracts between users within the same irrigation district. In Minle County, informal water exchanges began as early as the 1960s and 1970s. Farmers along the same canal traded privately on an informal basis. Occasionally village-level transactions took place across irrigation canal branches. Although at the time such transactions were considered illegal, the government did not impose restrictions on these exchanges, possibly due to their low frequency. The transactions only entailed water use within a growing season, and transaction prices were negotiated between buyers and sellers, at prices usually well above the fees charged by the irrigation districts. For example, when the irrigation districts charged a fee of 0.035 yuan/cubic metre, the transaction price could reach as high as 0.1 yuan/cubic metre.

In 1966, Mingle County Government stated that future water allocations, irrigation infrastructure construction and maintenance responsibilities would be based on the extent of irrigated land reported by each village. The reported numbers were then verified and confirmed by relevant government agencies and irrigation district officers, which became the initial water use right for each farmer. In Hongshuihe Irrigation District (HID), the total initial water use right defined in acreage was 9099 hectares in 1966, which were then assigned to each individual farmer and provided the basis for water use to be transferred among farmers. However, as each farmer's water use right was relatively small, and there existed no water exchange platform or legal documents to regulate water transfers, transfers between farmers were usually temporary and/or informal. For example, Farmer A may have an agreement with Farmer B that Farmer A will irrigate 50 per cent less water during the first scheduled irrigation of the season, and the remaining 50 per cent of water will be transferred to Farmer B. During a later scheduled irrigation, Farmer B will irrigate less and return the same amount of water to Farmer A. In such instances, the land belonging to Farmers A and B is generally located within the same channel. Although such water exchanges have become increasingly common, there is no record-keeping.

Formal water right transactions have taken place since 2015, including transactions at the village level and at the farm level. Currently applications for formal water transfers, for either one year or multiple years, can be made to the irrigation district management. Since the required minimum volume of a water transfer is 50 000 m³, much larger than the amount of water used by a typical farmer, trading bodies are usually WUAs from different villages (generally a village has one WUA). These two Water User Associations first negotiate on their own, sign a formal contract, and then seek the consent of the irrigation district to officially file the contract within that district. At present, only the HID has maintained record-keeping for such transactions. The interviewed irrigation district managers stated that one of the priorities in water management is to establish a trading platform.

There are currently only 18 cases of water trading recorded in the HID, with the majority of transactions being cross-irrigation channel transactions and cross-water user association transactions. There is currently no trade between the agriculture sector and industries. Neither are there any transactions at the county level or above. While the parties involved negotiate the price of the water, it cannot be higher than the limit set by the government, which is three times the fee charged by the irrigation district. The details on WUAs involved in water trading include what villages are not growing crops in particular timeframes, and the location to which their water use rights are transferred, for a given price. Applications are usually approved by the township governments of the villages and the irrigation district management.

Applying the first step of the WMRA framework, Zhangye City's reliance on surface water from the Heihe River and water allocation is a top-down process, with a total cap set for the whole basin and also separately for the upper, middle and lower reaches of the basin, which is distributed to city, county and irrigation district levels. In the Heihe River Basin, ShuiZiZheng [1997] No.496 (The Ministry of Water Resources) defines the water allocation plan for upper, middle and lower reaches in wet, normal and dry years, and over different seasons within the year. This water allocation plan ensures that water is not over-allocated to irrigated agriculture and that there is water available for the oasis area in the lower reaches, particularly during dry years.

Water allocated to Zhangye City from the middle reaches of the basin is distributed according to population and industry development (Sun et al., 2016). The amount of water allocated to agriculture is then distributed to each county in line with the status of its water resources, water rights area and cropping structure. Each county further allocates water to irrigation districts according to the same principle. Irrigation districts apportion agricultural water to each farm household based on their water rights area and irrigation quota. However, the interviews suggest that, in practice, this allocation structure at the county level plays a limited role. For example, in Minle County, the total water use cap (including agriculture, industry, and domestic water) has been set by the government at 417 million m³ in 2015, 364 million m³ in 2020, and 375 million m³ by 2030. These cap limits are not meaningful for Minle County given that, as a water-deficient area, its annual water availability is between 310 and 320 million m³, well below the three cap limits. In Gaotai County, the total water use cap (including agriculture, industry and domestic water) has been set by the government as 389 million m³ in 2015, 340 million m³ in 2020, and 350 million m³ in 2030. However, according to this plan, water for agriculture is well below the crop water requirement in this county, resulting in water consumption above the cap. However, there is no sanction mechanism for over-use, resulting in a non-binding role of the cap.

With regard to existing institutional, planning and property right arrange-ments, as outlined under the first step of the WMRA framework, a water rights system has been promoted in Zhangye City since 2002. Since rivers in Minle County do not flow into the central Heihe River, the county's water resources are a closed system. The water rights area is the basis for water allocation. In 2002 a water rights area for each household was established, and remained unchanged at 574,000 mu (38 267 ha), for the whole county. However, in 2017 the actual irrigated area reached 72 380 ha. Regarding the amount of water allocated to each unit of water rights area, Mingle County has adopted a fixed-amount quota approach. According to the difference between precip-itation and evaporation, the county is divided into three regions (region one with more evaporation relative to precipitation; region two with evaporation similar to precipitation; and region three with more precipitation relative to evaporation) with a different quota of water volume per irrigation round. Since 2002, during the crop-growing period, the quota has been set at 1.14 ML/ha/round for region one, 1.11–1.12 ML/ha/round for region two, and 1.05 ML/ha/round for region three.

The issuance of water right certificates also began in 2002. The certificates have different forms and can be water right certificates for irrigation districts, WUAs or rural households. Both Minle County and Gaotai County at present issue water right certificates at all three levels. In 2015, Mingle started to issue water right certificates to individual farmers, and every farmer should have been issued with their own water right certificate by the end of 2017. The Water Affairs Bureau manages all water allocation and water right transac-tions, and the water right certificate is the only document citing proof of water right ownership. The water right certificate includes the following information: year of issuance, land area, approved water right area and approved total water use indicator. The approved total water use indicator is a dynamic indicator, and depends on water availability in any given year. With the gradual rise of water right trading, water right certificates play an important role as the legal document for owning water rights.

Since the water rights area of Minle County is much smaller than the actual irrigated area, Minle County Water Affairs Bureau, along with various irrigation districts and other water supply departments, allocate water according to the proportion of water rights in each round of irrigation. In order to ensure that water is allocated to all farmers with water rights, the proportion is the same across upstream and downstream farms. Once the proportion of water allocation is determined, each farmer can calculate how much water they will receive for a given irrigation round. For example, if the water allocation proportion is 0.4 (due to the lack of water in Minle County, where the proportion is generally less than 1), and a farmer has 1 ha of irrigated land, but owns a water rights area of only 0.5 ha, according to the allocation proportion of 0.4 in this round, this farmer will only receive water for 0.2 ha of irrigated land (0.5 ha*0.4 = 0.2 ha). The farmer in this case will generally have three choices: (1) irrigate only 0.2 ha this round, and wait until the next round to irrigate the remaining land; (2) reduce the water application rate and irrigate as much land as possible; or (3) a combination of the first two. Farmers in Mingle County have already chosen a combination of the two options, and grow their crops so that not all of the irrigated area needs to be irrigated during each irrigation round. By adjusting the crop mix, farmers are able to stagger the irrigation needs of different plots. For example, maize does not generally require irrigation in the first two rounds, whereas wheat has already been harvested by the time of the final irrigation rounds. Different water needs among farmers also create opportunities for water to be traded or exchanged. As a result, it is not uncommon to observe farmers exchanging water during one irrigation round for water in another round.

In terms of the WMRA's second step of evaluating the potential benefits from trade, a number of aspects need to be considered: externalities, governance/institution costs, transaction costs, and number of water users and sectoral activity (Wheeler et al., 2017). Based on a few recorded transactions and undocumented exchanges among farmers, it seems that both buyers and sellers benefit from trading, given that sellers receive extra income from the sale while buyers secure additional water for their crop needs. However buyers and sellers, in the absence of regulation, do not generally consider third-party effects such as reliability of delivery, storage and delivery charges, and water quality (Heaney et al., 2006). Transaction price is usually negotiated privately between sellers and buyers, and there is no means to compel both parties to include negative third-party effects in their decision-making. In addition, governance and institutional costs for water trade are expected to be significant, considering that the market is currently under-developed. The government needs to invest a substantial amount of resources to develop market institutions, which may not be justified given the extremely low rate of water trade. Further, current irrigation infrastructure needs to be upgraded in order to make formal water trading at the individual level effective and efficient. For example, water metering at the individual plot level is not available and irrigation timing is still centrally controlled. For informal water exchange between farmers, transaction costs are low since there is only oral agreement and both parties are likely to have already known each other. However, for formal inter-sectoral or between-WUA trades, there are usually substantial transaction costs involved, such as searching for buyers and sellers, price negotiation, contract approval and signing. Given current small farm sizes and immature legal systems and institutional settings, trading among farmers (both groundwater and surface water) is likely to stay informal in the foreseeable future.

There are a large number of small-scale farmers, each with a small amount of water rights. The majority of farmers grow staple crops including grains, beans and potatoes – accounting for 65 per cent of total planted area. Other crops include oilseed (4 per cent), vegetables and melons (4 per cent), Chinese medicines (16.6 per cent) and orchard fruit (10 per cent) (Gansu Province Statistical Bureau, 2017). Flood irrigation is used by the majority of farmers (72.7 per cent of total irrigated area), while sprinkler, drip and low-pressure irrigation infrastructure accounts for 2 per cent, 5.4 per cent and 19.9 per cent of total irrigated area, respectively (Gansu Province Statistical Bureau, 2017). Consequently, although there appears to be a large number of heterogeneous water users with different crops and levels of water need, these water users may have only a small amount of water available for trade. Therefore, without an exchange platform and irrigation infrastructure to facilitate trade at the individual level efficiently and effectively, formal water trade between farmers is not likely to result in substantial benefits, and hence is not widely adopted. From the step two assessment, the conclusion is that potential benefits from formal water trade among farmers do not sufficiently outweigh costs at present. The WMRA framework suggests, in this case, maintaining the status quo, with enablers for trade and further monitoring if future demand or context changes.

4.4   CONCLUSION

Applying the WMRA framework (Wheeler et al., 2017) to the existing water market in Zhangye City, within the Heihe River Basin, this chapter assesses a number of market enablers and barriers including property rights/institutions, hydrology, externalities/governance, system type, adjustment, and entitlement registers and accounting processes. The framework covers a comprehensive list of assessing points for a functional water market and is generally applicable to the Chinese context, although it needs to be acknowledged that most farmers in China have relatively smaller amounts of water. Therefore, trades between farmers would appear inconsequential, and consequently farmers are not incentivised to engage in trades. With continuing urbanisation and rural land consolidation in China, irrigation farms are expected to increase in size and water is likely to become a substantial asset of an irrigation farm, as is the case in the Murray–Darling Basin in Australia (Seidl et al., 2020a). Therefore the assessment outcome for water markets in Zhangye City in the Heihe River Basin – maintaining the status quo, with enablers for trade and further monitoring if future demand or context changes – is appropriate.

While some of these components are assessed as providing good evidence to support water market readiness, there are many areas indicating limited evidence or further reform required, such as understanding trade impacts, and the monitoring and enforcement of water use. Importantly, trade impacts on stakeholders would depend on the size of each trade, and the cost of monitoring and enforcement of water use is likely to decrease as the number of farms reduces. Therefore, future urbanisation and rural land consolidation undoubtedly will accelerate progress towards markets for water.

The assessment suggests that net benefits can be realised by promoting water market development in the study area. Sun et al. (2016) found that the current low level of water trading activity in the Heihe River Basin resulted in only limited benefit, due to a number of challenges, including: ineffective implementation; the high cost of implementation given the large number of small-scale farmers; and lack of consistency in applying punitive measures to hold water users accountable. Moore (2014) stated that among these challenges that could be improved in implementation, perhaps the most important ones concern effective monitoring and enforcement of water use. Newer technologies, such as well-level monitors and networked streamflow, will reduce the currently high costs involved in tracking and verifying water use and thus hold irrigators more accountable. Detailed regulation requirements for trade and the availability of expert personnel for conducting trade, recording transactions and updating registers will enable market readiness as well. And finally, as transparency is critical to well-functioning markets (Seidl et al., 2020b), the establishment of a central location for trade and market information will assist market participants and stakeholders in making informed decisions.

ACKNOWLEDGEMENTS

This work was supported by the Australian Research Council [DP140103946 and DP200101191], an Australian National Commission for UNESCO grant – 'Developing a water market readiness framework', and the General programme of humanities and social sciences research of the Ministry of Education of China [19YJC790118].

NOTES

1. Piloted by the World Bank in 1995 in Hubei province, China (World Bank, 2003), WUAs are farmer-run, participatory institutions that are created to take the place of traditional, village leader-run water control organizations.
2. One ML (mega litre) is equal to 1 million litres, or 1000 cubic metres, or approximately 0.810 acre-feet. As of 2018, 1 RMB was equal to approximately US$0.15, resulting in a range between US$9 and US$57.75.

REFERENCES

Araral, E. (2010). Reform of water institutions: review of evidences and international experiences. *Water Policy*, *12*(S1), 8–22.

Aregay, F., Zhao, M., and Bhutta, Z. (2013). Irrigation water pricing policy for water demand and environmental management: a case study in the Weihe River basin. *Water Policy*, *15*(5), 816–829.

Calow, R., Howarth, S., and Wang, J. (2009). Irrigation development and water rights reform in China. *Water Resources Development*, *25*(2), 227–248.

Chong, H., and Sunding, D. (2006). Water markets and trading. *Annual Review of Environmental Resource Economics*, *31*, 239–264.

Debaere, P., Richter, B., Davis, K., Duvall, M., Gephart, J., et al. (2014). Water markets as a response to scarcity. *Water Policy*, *16*(4), 625–649.

Etchells, T., Malano, H., and McMahon, T. (2006). Overcoming third party effects from water trading in the Murray–Darling Basin. *Water Policy*, *8*, 69–80.

Gansu Province Statistical Bureau (2017). *Gansu Agriculture Yearbook*. China Statistics Press.

Gao, E. (2007). *China Water Rights System Development*. China Water Conservancy and Hydropower Press.

Gould, G. (1988). Water rights transfers and third-party effects. *Land and Water Law Review*, *23*(1), 1–43.

Grafton, Q., Libecap, G., Edwards, E., O'Brien, R., and Landry, C. (2012). Comparative assessment of water markets: insights from the Murray–Darling Basin of Australia and the western USA. *Water Policy*, *14*, 175–193.

Heaney, A., Dwyer, G., Beare, S., Peterson, D., and Pechev, L. (2006). Third-party effects of water trading and potential policy responses. *Australian Journal of Agricultural and Resource Economics*, *50*(3), 277–293.

Howe, C., Schurmeier, D., and Shaw, W. (1986). Innovative approaches to water allocation: the potential for water markets. *Water Resources Research*, *22*(4), 439–445.

Huang, Q., Rozelle, S., Howitt, R., Wang, J., and Huang, J. (2010). Irrigation water demand and implications for water pricing policy in rural China. *Environment and Development Economics*, *15*, 293–319.

Jiang, M. (2018). Recent development of water trading in China. *The Asia Dialogue*. https://theasiadialogue.com/2018/05/29/recent-developments-of-water-trading-in -china/.

Johansson, R., Tsur, Y., Roe, T., Doukkali, R., and Dinar, A. (2002). Pricing irrigation water: a review of theory and practice. *Water Policy*, *4*, 173–199.

Lewis, D., and Zheng, H. (2018). How could water markets like Australia's work in China? *International Journal of Water Resources Management*, *34*(3), 1–21.

Lui, Y. (2004). *Investigation on Problems Related to Water Rights Trading Markets*. China Water Conservancy International Cooperation and Science and Technology Net. http://www.cws.net.cn/jour- nal/cwr/200409/02.htm.

Moore, S. (2014). Water markets in China – challenges, opportunities, and constraints in the development of market-based mechanisms for water resource allocation in the People's Republic of China. Discussion Paper, Harvard University.

MWR (2016). *Temporary Management Regulation on Water Right Transfer*. Ministry of Water Resources, Beijing. http://www.gov.cn/zhengce/2016-05/22/content _5075679.htm.

Ray, D. (2013). Water trading could exacerbate water shortages in China. https://www .chinadialogue.net/article/show/single/en/6026-Water-trading-could-exacerbate -water-shortages-in-China.

Seidl, C., Wheeler, S.A., and Zuo, A. (2020a). High turbidity: water valuation and accounting in the Murray–Darling Basin. *Agricultural Water Management*, *230*, 105929. https://doi.org/10.1016/j.agwat.2019.105929

Seidl, C., Wheeler, S.A., and Zuo, A. (2020b). Treating water markets like stock markets: key water market reform lessons in the Murray–Darling Basin. *Journal of Hydrology*, *581*, 124399. https://doi.org/10.1016/j.jhydrol.2019.124399.

Speed, R. (2009). A comparison of water rights systems in China and Australia. *International Journal of Water Resources Development*, *25*(2), 389–405.

Sun, T., Wang, J., Huang, Q., and Li, Y. (2016). Assessment of water rights and irriga- tion pricing reforms in Heihe River Basin in China. *Water*, *8*(8), 333–348.

Sun, X. (2009). The development of a water rights system in China. *International Journal of Water Resources Development*, *25*(2), 189–192.

Wang, J., Jiang, Y., Wang, H., Huang, Q., and Deng, H. (2019a). Groundwater irriga- tion and management in northern China: status, trends, and challenges. *International Journal of Water Resources Development*, *36*(4), 1–27.

Wang, J., Li, Y., Huang, J., Yan, T., and Sun, T. (2017). Growing water scarcity, food security and government responses in China. *Global Food Security*, *14*, 9–17.

Wang, J., Zhang, L., and Huang, J. (2016). How could we realise a win–win strategy on irrigation price reform: evalution of a pilot reform project in Hebei Province. *China Journal of Hydrology*, *539*, 379–391.

Wang, J., Zhu, Y., Sun, T., Huang, J., Zhang, L., et al. (2019b). Forty years of irrigation development and reform in China. *Australian Journal of Agricultural and Resource Economics*, *64*(1), 126–149.

Wheeler, S.A., Loch, A., Crase, L., Young, M., and Grafton, Q. (2017). Developing a water market readiness assessment framework. *Journal of Hydrology*, *552*, 807–820.

World Bank (2003). *Water User Association Development in China: Participatory Management Practice under Bank-supported Projects and Beyond*. Social

Development Notes, No. 83, June. http://documents.vsemirnyjbank.org/curated/ru/ 431861468259759333/pdf/279930BRI0Repl1ame0catalogind0info1.pdf.

Xia, C., and Pahl-Wostl, C. (2012). The process of innovation during transition to a water-saving society in China. *Water Policy, 14,* 447–469.

Xie, J. (2009). *Addressing China's Water Scarcity.* World Bank.

Zhong, L., and Mol, A. (2010). Water price reforms in China: policy-making and implementation. *Water Resources Management, 24,* 377–396.

Zhu, Z., Zhou, H., Ouyang, T., Deng, Q., Kuang, Y., and Huang, N. (2001). Water shortage: a serious problem in sustainable development of China. *International Journal of Sustainable Development and World Ecology, 8* (3), 233–237.

5. When the genie is out of the bottle: the case of dynamic groundwater markets in West Bengal, India

Sophie Lountain, Lin Crase and Bethany Cooper

5.1 INTRODUCTION

Water markets have long been touted as a means of resolving conflicts over water scarcity and acting as a vehicle to allocate water resources efficiently over time. The elegance of the market mechanism is compelling, at least at a theoretical level, such that proponents of water markets are seldom questioned on academic grounds. However, the practicality of establishing such markets from scratch can be challenging for legislators; a point emphasised by the water market readiness assessment framework analysis (Wheeler et al., 2017).

The market readiness assessment framework offers a practical guide to those seeking to develop water markets and thereby address emerging concerns about water scarcity. The authors of the guide note that 'market-based reallocation may be unsuitable for developing countries' (Wheeler et al., 2017, p. 808), highlighting challenges with the definition of water rights, poor infrastructure and data to manage water and weaknesses in the rule of law, amongst others. However, it is not clear how the framework could be used in circumstances where water markets have already emerged in developing countries, and continue to evolve in response to the changing administrative and policy environments that circumscribe them.

These types of markets can be found in South Asia – India in particular – and especially in the context of groundwater. Groundwater markets have been analysed in India extensively. However, gaining an appreciation of their functioning is complicated by the wide range of approaches to water management that occur across the different states. Nonetheless, a greater understanding of these markets has much to offer, given the magnitude of the population now reliant on agriculture and groundwater extraction for their livelihoods.

This chapter offers insights into the functioning of existing groundwater markets in India. Whilst some national perspectives are presented, the bulk of the chapter centres on West Bengal, where groundwater markets stand to make a significant difference to the well-being of many agrarian households. We trace the evolution of those markets and consider the complex interactions that are occurring as governments modify policies and incentives relating to different pumping technologies. We then use this analysis to consider how the market readiness assessment framework might be reconfigured to offer benefits to policy-makers dealing with such a dynamic and challenging landscape.

5.2   HISTORY OF GROUNDWATER USE IN INDIA

5.2.1   Groundwater Irrigation and Ownership

After the British colonisation of India, the country's traditional systems of water collection and distribution were abandoned, and the ownership of natural resources was transferred to the British. Following the 1857 Revolution, control over surface waters by the state increased, leading to significant investment in canals and irrigation facilities.

Later, following the Transfer of Property Act, 1882,[1] landowners' rights to water were legitimised based on the 'dominant heritage' principle. The Indian Easements Act, 1882[2] assigned private rights over groundwater by considering it as connected to the land above it. This amounted to a system of corresponding rights, where an individual's claim to groundwater was proportionate to their land ownership. Although India is now an independent country, able to make its own laws, water and land rights remain linked. As a result of weak legal provisions, access to groundwater runs on a system where farmers with larger landholdings, greater pumping capacity and deeper wells have the most control over the resource (Saleth, 1998).

The responsibility of water was moved to be under the jurisdiction of the state governments with the Government of India Act, 1935[3] and with the Green Revolution came the beginning of intensive irrigation development across the country. The late 1970s saw Indian states provide subsidised and unmetered electricity to farmers for agricultural purposes, leading to an immense rise in the use of groundwater for irrigation throughout the 1970s and 1980s (Mukherji and Das, 2014). According to the agricultural censuses, in the ten years between 1980–81 and 1990–91, the area irrigated by tubewells increased fivefold (Acharyya et al., 2018). Other estimates report that from 1980 the expansion was about 23 per cent per annum (Rawal, 2001). As a result of low levels of interest from the government, tubewell expansion mainly came as a result of private investment by millions of individual farmers across the country (Acharyya, 2013, as cited in Acharyya et al., 2018). This growth was especially apparent in West Bengal. It was around this time that West Bengal saw an increase in crop production, especially of summer (*boro*) paddy (Acharyya et al., 2018).

5.2.2 Introduction to India's Groundwater Market

Groundwater irrigation is considered to be the most productive source of irrigation in India, and it is relied upon heavily to support its predominantly agrarian economies. In the early 2000s, groundwater irrigation made up close to 60 per cent of India's irrigated area of 56 million hectares (Government of India, 2003, as cited in Mukherji, 2007), and it provided more water to agriculture than all other sources of irrigation combined (Deb Roy and Shah, 2002).

India's agricultural population is primarily comprised of small and marginal farmers. Many of these farmers are not financially able to own personal tube-wells, and accordingly access water through private purchase from wealthier pump owners in informal groundwater markets. As of 1999, 21 million households in India owned water extraction mechanisms (WEMs). In comparison, 24 million households hired irrigation services from others (NSSO, 1999). Through enabling access to groundwater irrigation, informal water markets can be considered as an 'important institutional mechanism' (Mukherjee and Biswas, 2016). Arguably, they improve equity by extending groundwater access across farmers of different financial standing, thereby increasing food security, agricultural productivity and poverty alleviation (Ananda and Aheeyar, 2019; Mukherjee and Biswas, 2016). Farmers who buy water or rent irrigation equipment through these informal markets can achieve similar cropping intensity, agricultural productivity and incomes as those who have their own pumps (Mukherji, 2007), and enjoy greater access to capital.

5.3 STATE-SPECIFIC CONSIDERATIONS AND THE ENERGY–IRRIGATION NEXUS

5.3.1 Energy–Irrigation Nexus and the Groundwater Market

The rate of development of informal water markets depends on a range of influences, including the availability of water resources, the quality of irrigation technologies and the scale of their adoption, the extent of rural electrification, quality of supply and cost, and the degree of land fragmentation (Shah, 1991). West Bengal, in particular, has extensive groundwater reserves, and yet under-developed water markets. Hindrances here are primarily from the supply side; groundwater irrigation is not fully developed, the land is highly fragmented, and the pace of rural electrification has been low (Acharyya et al., 2018; Daschowdhury et al., 2009; Saleth, 1998; Shah, 1991).

The energy and irrigation sectors form a significant relationship in groundwater markets. This relationship is often known as the 'energy–irrigation nexus' (Bassi, 2015; Beaton et al., 2019; Daschowdhury et al., 2009; Mukherji, 2007; Shah et al., 2003), and it is important because the way pumps are powered, and the cost of doing so, dramatically affects how these markets function (Shah et al., 2003). Water for irrigation in India is accessed using a variety of different WEMs powered by liquid fossil fuels (that is, diesel and kerosene), grid electricity or electricity from solar panels, or a combination. According to the 5th Census of Minor Irrigation Schemes report, in 2014/15, there were 21.7 million irrigation schemes in India, of which 94.5 per cent drew on groundwater. Notably, 20.2 million of these had a water-lifting device; 72 per cent were powered by grid electricity, and 23.7 per cent by diesel. Only 2874 pumps (0.01 per cent) were powered solely by solar energy; 10 112 (0.05 per cent) ran on a combination of solar and grid electricity, and 2270 (0.01 per cent) on a mix of solar and diesel (Government of India, 2017). Clearly, lifting water using electricity is a significant issue.

There are generally three ways in which electric WEM owners pay for electricity (Shah, 1991). Firstly, there is a pro rata tariff, which is based on meter readings where each farmer pays for the power used. Secondly, a flat, horsepower-linked fare, where a farmer pays a flat rate regardless of the actual energy used. The final option is a combination of the two, where farmers pay a part of the power bill as a fixed charge and the remainder as a metered charge. Obviously, and in contrast, where owners of diesel WEMs are purchasing fuel as needed this ostensibly amounts to facing a fixed charge for access to the machine and a variable charge for usage.

When water is sold for cash in the informal market, pricing is mostly dictated by the type of pump used and the buyer's use (Saleth, 1998). For instance, while some electric WEM owners may demand hourly rates, area-based rates are also common. Here, the price is dependent on the type of crop irrigated, or the time of the season. However, while hourly or area-based rates are given to regular water buyers, different charges are assigned to sporadic purchases for supplementary irrigation. In terms of payment, for farmers purchasing for the month or season, credit is usually permitted for electric pump use, but for occasional water sales, immediate payment is required. Use of diesel pumps also requires immediate payment, and the price is based on an hourly rate, as diesel consumption is associated with hours of operation (Saleth, 1998).

5.3.2   Impact of the Electricity Subsidy

In recent years, the government of India has worked towards improving electrification in India's rural areas (Heynen et al., 2019). As a result of this initiative, 100 per cent of all villages are purportedly now considered to be connected to electricity (D'Cunha, 2018). However, this number is not entirely reflective of individual access: a quarter of the population still has no direct electricity access, and 40 per cent only have partial access (Winkler et al., 2018).

In the 1970s, the Indian government introduced an agricultural electricity subsidy, whereby farmers were supplied with unmetered electricity. Previously, all state electricity boards charged for electricity based on metred tariffs. However, as the number of tubewells increased during the 1970s and 1980s, the transaction costs of metering became excessive, and the government introduced a flat-rate tariff (Shah et al., 2007). Initially, the flat-rate tariff was to remain in sync with the cost of electricity generation and supply (Mukherji et al., 2009). However, it soon became clear what a valuable tool a low rate was in garnering electoral support, so the cost remained modest in most states (Dubash and Rajan, 2001).

From an economic perspective, supplying electricity on a flat-rate tariff clearly affects the behaviour of water sellers by reducing the incremental pumping costs to practically zero. This also has implications for the monopoly power of sellers (Shah, 1991). These effects occur because, under a flat-rate tariff, sellers experience a natural motivation to increase their pump use by selling larger amounts of water to increase their profits. This incentive boosts competition among sellers and forces a decrease in the market price of water. This situation generates a surplus for some; more specifically, it allows more resource-poor farmers to access irrigation.

5.3.3 Western India

Generally, water markets are more developed in the west of India. Here, the groundwater irrigation economy is dominated by electric pumps, which run on free or subsidised electricity. Power that is free or highly subsidised means that the marginal cost of water extraction for farmers who own pumps is close to zero (Mukherjee and Biswas, 2016). This minimal cost encourages groundwater use and has facilitated competitive water markets among pump owners. For buyers, the cost of water is generally consistent across sellers and close to the marginal cost of pumping. In this part of the country, it is not a significant limitation to be a water buyer, due to the price and many opportunities for purchase (Shah, 1991).

In these drought-prone states, however, there is growing concern over groundwater over-exploitation. Electricity subsidies have lasting impacts on the sustainability of groundwater resources, and the unavailability of groundwater can have dire implications for farmers, especially poorer farmers who are not able to transition out of the agricultural sector (Mukherjee and Biswas, 2016). Concerns over this, and the increasing cost of subsidising electricity, has led to the search for a balance between irrigation, agricultural productivity and what Shah and Chowdhury (2017) describe as 'socioecological sustainability'.

5.3.4   Eastern India

A different situation exists in India's eastern states. In general, this part of the country is considered abundant with groundwater resources, but it is also characterised by limited rural electrification and high diesel prices which constrain the development of groundwater irrigation (Mukherji, 2007). Until the late 1990s, the government subsidised the cost of diesel prices; the cessation of these subsidies increased the shelf price of diesel and has led to an 'economic scarcity of groundwater' (Mukherji, 2007), where people are unable to access groundwater due to prohibitive extraction costs. In this situation, water sellers enjoy monopoly power and charge higher prices. While this condition can result in higher profits for sellers, the impacts on market efficiency are questionable. Moreover, this condition has led to a severe recession in the informal groundwater markets, risking the livelihoods of millions of subsistence farmers (Mukherji, 2007).

5.4   WEST BENGAL

5.4.1   Agricultural Water Use in West Bengal

As noted earlier, the conditions that circumscribe groundwater markets in the east and west of India vary considerably. West Bengal is an eastern state of India, with a relatively high annual rainfall (1500–2500 millimetres on average) and abundant surface water resources. An estimation performed by the State Water Investigation Directorate and the Central Groundwater Board revealed that of the 269 blocks in the state, 86 per cent (231 blocks) were declared 'safe' in terms of water volume (Mukherji, 2007). Thirty-seven blocks were deemed 'semi-critical', and only one block was considered 'critical'.

As a state, West Bengal produces the largest volume of rice and vegetables in the country. There are more than 7 million farm families in the state, of which 96 per cent are small and marginal farmers with an average landholding of 0.77 hectares (Mukherji, 2007). Farmers in West Bengal must grow crops during both the wet and dry seasons to earn a livelihood. However, water reserves are usually dry by January, and this persists until the monsoon rains start half a year later. While groundwater resources are sufficient in West Bengal and using groundwater for irrigation is necessary for adequate agricultural production, in many instances state polices have tended towards discouraging farmers from doing so (Daschowdhury et al., 2009).

5.4.2 Groundwater Markets in West Bengal

The rise of informal groundwater markets began in West Bengal in the 1980s when the government introduced a flat-rate electricity tariff for farmers using tubewell irrigation (Mukherji, 2007). This regime produced a buyers' water market, where owners of electric submersible pumps were under constant pressure to generate income to pay their high bills. Suppliers of groundwater (that is, owners of electric pumps) competed vehemently to sell, offering farmers top irrigation services for a low price, compared with diesel shallow tubewell owners.

Despite the flat tariff remaining low in other Indian states, in West Bengal the costs progressively increased from the early 1990s (Mukherji and Das, 2014). For example, the cost of electricity in West Bengal was set at Rs 0.92/unit in 2005. In comparison, the price was only Rs 0.42/unit in Haryana and Rs 0.62/unit in Gujarat (Narendranath et al., 2005). During this time, tubewell owners in West Bengal absorbed the majority of tariff increases, for fear of losing their market share, leading to consistent losses on an individual level. When the flat-rate tariff increased sixfold from Rs 1100 per year to Rs 6850 per year for submersible pumps, water rates only increased twofold, from Rs 625 per acre to Rs 1500 per for summer (*boro*) paddy (Mukherji et al., 2009).

5.4.3 Groundwater Policies

In 2007, the government of West Bengal introduced a significant groundwater policy reform: the reintroduction of universal metering of electric tubewells. This shift to metering came as a result of the collective demands of local tubewell owners, and a national policy focused on reducing losses for the state electricity boards and improving the sustainability of groundwater use across the country (Jha, 2017).

With this policy change, also came the requirement that farmers apply for a permit to sink a well. Accessing a permit was a complicated, lengthy and expensive process, and smaller farmers were effectively priced out of owning a WEM. Farmers also had to pay the full cost of connecting their pump to the electricity grid, including the purchase and installation of wires, poles and transformers. The total cost of this was based on installation distance from the network. Accordingly, farmers who lived far from the network found this extremely difficult to afford (Mukherji et al., 2012; Shah and Chowdhury, 2017).

West Bengal is one of the few Indian states to re-implement metering, despite a low reliance on agricultural electricity due to poor reliability and access. While the national average for electrification was 51 per cent, just over 10 per cent of pumps in West Bengal were powered by electricity (NSSO,

1999). Under metering, the cost of electricity follows a time-of-day pricing system, which charges users different rates for the volume of water pumped, depending on the time of use (Mukherji, 2007).

Metering had a profound impact on the dynamics of the groundwater market. A year after metering took place, farmers engaged in the groundwater market were surveyed about the change. Of the farmers surveyed, two-thirds of pump owners preferred the metered tariff, while only a quarter of the water buyers supported this approach (Mukherji et al., 2009). While metering had resulted in lower water costs for pump owners (wealthier farmers), these cost savings were not passed on to water buyers (poorer farmers), and the price of water for this group had increased as much as 50 per cent (Mukherji et al., 2009).

Under metering, WEM owners were under less pressure to sell their water, and the buyers' water market quickly reverted to one favouring sellers. While this shift benefited some WEM owners and the state electricity board, the change was likely unfavourable for millions of small and marginal water buyers (Jha, 2017).

In light of this information, in November 2011 and 2012, the government of West Bengal altered the groundwater legislation (Mukherji, 2007). The aim was to remove barriers to accessing groundwater, sustainably. The first change was the West Bengal Groundwater Resources Act 2005, which was modified to allow farmers in specific areas to install low-powered pumps without applying for a permit. This revision made it easier for farmers in areas with potential for groundwater development to access groundwater.

The second change came in 2012 and involved altering electricity policy and procedures. A new scheme was launched, which allowed farmers to access electricity connections at a fixed cost. This meant that instead of paying the total cost for equipment and installation based on how far they were from a connection, the price was set at a specified amount. For some farmers in West Bengal, this made accessing an electricity connection for irrigation far more affordable.

In Eastern India, and West Bengal in particular, the development of water markets has the potential to generate vast improvements to the lives of poor farmers (Shah, 1991). Still, investments in rural electrification are essential to help increase groundwater use and boost agricultural productivity. How this occurs without causing stresses on other fronts remains a significant challenge.

5.5   APPLYING THE MARKET READINESS ASSESSMENT FRAMEWORK IN THIS CONTEXT AND CONCLUDING REMARKS

Clearly, the environment in which groundwater is traded in West Bengal is dynamic. What is also clear is the pivotal link between water trade and adjustments in the way energy is sourced and paid for. In contrast, the framework posited by Wheeler et al. (2017) focuses heavily on the institutions that relate to water, and the capacity of governments to develop the conditions that would be conducive to water trade. The development of the framework rests heavily on the experiences in Australia's Murray–Darling Basin, where water trade (particularly in surface water) has become a prominent feature of the agricultural landscape. In addition, the examples where the framework was tested included the United States, Spain and Tasmania in Australia. In each of these examples, the three steps of the water market readiness framework were assessed, and we endeavour to apply a similar analysis in the case of West Bengal groundwater.

5.5.1   Step 1: Background Context

The two components here relate to hydrology and institutions. In the case of the former, West Bengal enjoys ample groundwater resources, and farmers have come to rely on its use to increase agricultural intensity. As noted earlier, almost 90 per cent of the blocks in the state are considered safe for additional exploitation. This relates to the rapid recharge that occurs from the monsoon and the inundation that typifies large parts of the east Gangetic plain. However, water stress is evident in the dry season and scarcity is sufficiently common to warrant attention, and markets can thus play a part.

In terms of the institutional setting of the status quo, farmers have clear property rights to access groundwater, at least in a legal sense. But access and use are contingent on access to energy. This is the major form of property right attenuation. West Bengal, unlike other states in India, also has considerable public capacity to manage the distribution of energy and deal with its pricing. In developing the readiness framework, Wheeler et al. (2017) have focused almost exclusively on the nature of property rights in water. What this example illustrates is that water property rights can be severely impacted upon by rights to complementary resources, and this, in turn, can severely hamper the formal development of water markets. Put differently, in West Bengal farmers *prima facie* appear to be trading water, but in practice they are trading access to energy to lift water. It is not clear how the readiness framework deals with these complexities.

5.5.2   Step 2: Market Development, Evaluation and Implementation

In this step, the gains from trade need to be evaluated and enumerated. They also need to be of sufficient scale to warrant progressing to the formalisation of the market. In the case of West Bengal, there are manifest gains from trade. Moreover, as we have demonstrated, in some instances trade has been key to providing access for many poorer farmers with insufficient capital to control a pumping device independently. However, in this case trade has emerged organically and has not involved direct actions by the state. Instead, the state's actions on energy have had indirect but significant impacts on how water is traded. The point is that the application of the market readiness framework in its current form is somewhat redundant. Moreover, the costs that would need to be borne by the state to formalise and monitor these organic markets are likely to be high and, given the millions of smallholder farmers, the net welfare from trying to formalise such markets makes progression to step 3 unnecessary.

There is little doubt that the readiness framework is a useful tool in some settings, but its helpfulness in cases like West Bengal is questionable. The application in West Bengal highlights one oversight in the development of the framework. More specifically, the institutions that attend complementary resources to water can seriously impact upon the development and functioning of water markets. The framework itself was informed by changes in Australia's Murray–Darling Basin, and economists frequently marvel at the subsequent depth and breadth that has emerged in the water markets in that Basin. However, what is often overlooked is the changes that occurred to complementary resources in Australian agriculture at the time of the formation of those water markets. Crase et al. (2015) note that significant land reforms occurred in tandem with the development of water markets in the Murray–Darling Basin. As late as the mid-1980s, access to land in state-run irrigation projects was severely attenuated. For instance, no farmer could aggregate land beyond a state-determined limit. In addition, women married to men who owned the equivalent of a 'home maintenance area' were prevented from owning land in their own name. The dismantling of these (now seemingly arcane) institutions undoubtedly helped to fuel the formal development of water markets.

In the case of West Bengal, we have argued that the institutions that circumscribe access and use of energy are pivotal to the way water is marketed. Considering water in isolation from other resources runs the risk of overlooking key factors that explain the dynamics of markets, and might also foster government intervention when it is either unwarranted or likely to be ineffective. Hopefully, this is a point considered by the Australian government as it offers advice to neighbouring countries on water management practices.

ACKNOWLEDGEMENTS

The authors acknowledge the support of the Australian Centre for International Agricultural Research [LWR/2018/104].

NOTES

1. The Transfer of Property Act, 1882: www.indiankanoon.org/doc/515323/.
2. Indian Easements Act, 1882: www.indiacode.nic.in/handle/123456789/2349.
3. Government of India Act, 1935: www.legislation.gov.uk/ukpga/1935/2/pdfs/ ukpga_19350002_en.pdf.

REFERENCES

Acharyya, A., Ghosh, M., and Bhattacharya, R. (2018). Groundwater market in West Bengal, India: does it display monopoly power? *Studies in Microeconomics, 6*(1–2), 105–129.

Ananda, J., and Aheeyar, M. (2019). An evaluation of groundwater institutions in India: a property rights perspective. *Environment, Development and Sustainability*, 1–19. doi:10.1007/s10668-019-00448-8.

Bassi, N. (2015). Irrigation and energy nexus. *Economic and Political Weekly, 50*(10), 63–66.

Beaton, C., Jain, P., Govindan, M., Garg, V., Murali, R., et al. (2019). *Mapping Policy for Solar Irrigation Across the Water–Energy–Food (WEF) Nexus in India.* www .iisd.org/sites/default/files/publications/solar-irrigation-across-wef-nexus-india.pdf.

Crase, L., O'Keefe, S., Wheeler, S.A., and Kinoshita, Y. (2015). Water trading in Australia: understanding the role of policy and serendipity. In K. Burnett, R. Howitt, J.A. Roumasset and C.A. Wada (eds), *Handbook of Water Economics and Institutions* (pp. 296–313). Routledge.

Daschowdhury, S., Banerjee, P., and Mukherji, A. (2009). Managing the energy–irrigation nexus in West Bengal, India. In B. Sharma, K. Villholth, A. Mukherji and J. Wang (eds), *Groundwater Governance in the Indo-Gangetic and Yellow River Basins* (pp. 279–292). CRC Press.

D'Cunha, S. (2018). Modi announces '100% village electrification', but 31 million Indian homes are still in the dark. www.forbes.com/sites/suparnadutt/2018/05/07/ modi-announces-100-village-electrification-but-31-million-homes-are-still-in-the -dark/.

Deb Roy, A., and Shah, T. (2002). Socio-ecology of groundwater irrigation in India, IMWI-TATA Water Policy Research Program Annual Partners' Meet, 2002. No. H029653, IMWI Working Papers, International Water Management Institute. www .EconPapers.repec.org/RePEc:iwt:worppr:h029653.

Dubash, N., and Rajan, S. (2001). Power politics: process of power sector reform in India. *Economic and Political Weekly, 36*(35), 3367–3390.

Government of India (2017). *Report of 5th Census of Minor Irrigation Schemes.* www .indiaenvironmentportal.org.in/files/file/Report%20of%205th%20Census%20of %20Minor%20Irrigation%20Schemes.pdf.

Heynen, A., Lant, P., Smart, S., Sridharan, S., and Greig, C. (2019). Off-grid opportunities and threats in the wake of India's electrification push. *Energy, Sustainability and Society, 9* (16), 95–103.

Jha, K. (2017). *Evidence-Informed Policy Change: Improved Access to Groundwater in West Bengal, India.* 3ie Evidence Use Brief Series, International Initiative for Impact evaluation (3ie). www.3ieimpact.org/sites/default/files/2018-12/evidence -impact-brief-wb-web.pdf.

Mukherjee, S., and Biswas, D. (2016). An enquiry into equity impact of groundwater markets in the context of subsidised energy pricing: a case study. *IIM Kozhikode Society and Management Review*, *5*(1), 63–73.

Mukherji, A. (2007). The energy–irrigation nexus and its impact on groundwater markets in eastern Indo-Gangetic basin: evidence from West Bengal, India. *Energy Policy*, *35*(12), 6413–6430.

Mukherji, A., and Das, A. (2014). The political economy of metering agricultural tube wells in West Bengal, India. *Water International*, *39*(5), 671–685.

Mukherji, A., Das, B., Majumdar, N., Nayak, N., Sethi, R., and Sharma, B. (2009). Metering of agricultural power supply in West Bengal, India: who gains and who loses? *Energy Policy*, *37*(12), 5530–5539.

Mukherji, A., Shah, T., and Banerjee, P. (2012). Kick-starting a Second Green Revolution in Bengal. *Economic and Political Weekly*, *47*(18), 27–30.

Narendranath, G., Shankari, U., and Rajendra Reddy, K. (2005). To free or not to free power: understanding the context of free power to agriculture. *Economic and Political Weekly*, *40*(53), 11.

National Sample Survey Organisation (NSSO) (1999). *54th Round: Cultivation Practices in India, January 1998–June 1998*. Department of Statistics and Programme Implementation, Government of India, August 1999. www.ilo.org/surveydata/index.php/catalog/145/study-description.

Rawal, V. (2001). Irrigation statistics in West Bengal. *Economic and Political Weekly*, *36*(27), 2537–2544.

Saleth, R.M. (1998). Water markets in India: economic and institutional aspects. In K.W. Easter, M.W. Rosegrant and A. Dinar (eds), *Markets for Water: Potential and Performance* (pp. 187–205). Springer US.

Shah, T. (1991). Water markets and irrigation development in India. *Indian Journal of Agricultural Economics*, *46*(3), 335–348.

Shah, T., and Chowdhury, S. (2017). Farm power policies and groundwater markets. *Economic and Political Weekly*, *52*(25–26), 29–47.

Shah, T., Christopher, S., Kishore, A., and Sharma, A. (2007). Energy–irrigation nexus in South Asia: improving groundwater conservation and power sector viability. In M. Giordano and K.G. Villholth (eds), *The Agricultural Groundwater Revolution: Opportunities and Threats to Development* (pp. 211–242). CABI Publishing.

Shah, T., Roy, A.D., Qureshi, A.S., and Wang, J. (2003), Sustaining Asia's ground-water boom: an overview of issues and evidence. *Natural Resources Forum*, *27*, 130–141. https://doi.org/10.1111/1477-8947.00048

Wheeler, S.A., Loch, A., Crase, L., Young, M., and Grafton, Q. (2017). Developing a water market readiness assessment framework. *Journal of Hydrology*, *552*, 807–820.

Winkler, B., Lewandowski, I., Voss, A., and Lemke, S. (2018). Transition towards renewable energy production? Potential in smallholder agricultural systems in West Bengal, India. *Sustainability*, *10*(3), 801.

6. Are water markets a viable proposition in the Lower Mekong Basin?

Kate Reardon-Smith, Matthew McCartney and Lisa-Maria Rebelo

6.1 INTRODUCTION AND BACKGROUND

Water security – access to water of suitable quality for livelihoods, development and ecosystem health, and protection against water-related natural disasters, social disruption and geopolitical conflict (United Nations University, 2013) – is a fundamental, but increasingly at-risk, human right (Harris et al., 2015). In many parts of the world, competition from diverse sectors for limited water resources requires careful management to ensure the equitable, efficient and sustainable use of available water. This necessitates sound knowledge and understanding of the hydrological system in question, appropriate governance and regulatory measures, and the institutional and community capacity to implement these.

Water scarcity requires both demand- and supply-side water resource management. On the demand side, water markets have been proposed and implemented in some countries as an effective financial mechanism by which, in theory, the most cost-efficient use of water can be achieved in water-limited environments. However, as Wheeler et al. (2017) point out, water market development will not be an appropriate solution in all contexts, and is likely to be especially difficult in transboundary basins comprising nations with divergent political systems and economies. Their proposed water market readiness assessment (WMRA) framework enables the identification of potential gaps in key policy and governance areas which might, unless addressed, constrain development of an effective water trading system. In the broader context of water resource management, the framework also provides a checklist for water resource knowledge and governance systems which are likely to be of benefit to greater water security overall.

In this chapter, we apply the WMRA framework within the Mekong River Basin (MRB) of mainland Southeast Asia, a region facing a number of complex geopolitical, socio-economic and environmental challenges. The

World Bank (WB), which along with the Asian Development Bank (ADB) funds major water development projects and has significant influence in developing countries, acknowledges that 'economic growth is a thirsty business', placing significant pressure on available water resources, infrastructure and governance (World Bank, 2019a). While water markets may play a role in building a more sustainable future in the region (Richter, 2016), a recent ADB report also acknowledges a range of policy, institutional and practical shortfalls in the application of market-based instruments (MBIs) such as water markets in such countries (Grainger et al., 2019). The WMRA framework systematically captures these and, in turn, may provide an evidence base for future investment strategies to enhance regional water security, whether or not water markets are seen as part of the solution.

6.2   CASE STUDY AREA DESCRIPTION

The Mekong is a large transboundary river system draining almost 40 per cent of mainland Southeast Asia (total length 4800 km; total catchment area 795 000 km²). Rising in China's Yunnan Province, near Tibet, it extends for 2200 km through southern China (the Upper Mekong) before flowing through five downstream Lower Mekong Basin (LMB) countries – Myanmar, Lao PDR, Thailand, Cambodia and Viet Nam – and into the South China Sea via the Mekong Delta in southern Viet Nam (Figure 6.1). In total, the LMB countries occupy almost 80 per cent of the total area of the Mekong River Basin (Table 6.1).

An estimated 65 million people live within the LMB (MRC, 2018b), the majority of these in rural areas, with many directly dependent on the river system for their livelihoods and well-being (Johns et al., 2010). For example, as well as the world's largest inland fishery (Hecht et al., 2019), LMB riverine ecosystems support significant rural economies which underpin the food and livelihood security of many millions of people (Johns et al., 2010; Hecht et al., 2019).

6.3   WATER RESOURCES AND THEIR DEVELOPMENT IN THE MEKONG BASIN

6.3.1   Water Resources

The MRB has significant water resources. In comparison with other international river basins, average annual discharge from the Mekong (totalling approximately 475 km³ or 13 000 m³/s) ranks eighth-largest, and per capita surface water resources are also high (FAO, 2011). Despite this, the Basin is

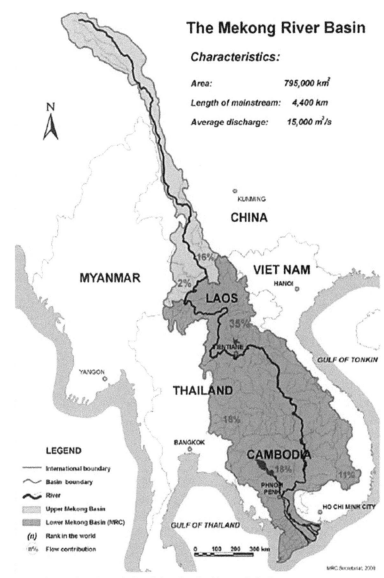

The Mekong River Basin

Characteristics:

Area: 795,000 km²

Length of mainstream: 4,400 km

Average discharge: 15,000 m³/s

Source: Mekong River Commission © (reprinted with permission).

Figure 6.1   *Mekong River Basin showing the relative contributions (%)*
to total Mekong River flow from constituent countries

Table 6.1   Countries situated in the Mekong River Basin

Sub-Basin	Country	Area within the MRB (km²)	% of total MRB area	% of total country area
Upper Mekong				
	China	165 000	21	2
Lower Mekong				
	Myanmar	24 000	3	4
	Lao PDR	202 000	25	85
	Thailand	184 000	23	36
	Cambodia	155 000	20	86
	Viet Nam	65 000	8	20

Note: Myanmar is not a signatory country to the MRC and is generally not included in MRC statistics.
Source: Adapted from FAO (2011).

subject to significant regional (spatial and temporal) variation in water availability (MRC, 2016a).

The climate of the Basin – hence streamflow in the river and its tributaries – is strongly affected by tropical storms and cyclones, in large part associated with the southwest monsoon (MRC, 2016a). This results in a strong annual pattern of water availability, governed by distinct wet and dry seasons, the intensity and duration of which vary under the influence of major climate drivers such as the El Niño Southern Oscillation (Villafuerte and Matsumoto, 2015; Ge et al., 2017). For example, Mekong River levels were at record low levels in June 2019 due to a combination of drought, a late start to the monsoon season, and upstream hydropolitics (MRC, 2019a; *National Geographic*, 2019). While drought is not commonly considered a significant risk in tropical monsoon regions (Adamson and Bird, 2010), the significant seasonality of rainfall can create critical water shortages for both agricultural and ecological systems in the dry season (Cook et al., 2012). Further increase in the variability of the region's climate is expected with global climate change, with projections of more intense rainfall (and flooding) during the wet season, and higher temperatures and increased evaporative stress at other times (Eastham et al., 2008; Gebretsadik et al., 2012).

6.3.2   Water Resource Development

Over recent decades, a combination of demographic, economic, political and cultural factors has driven (and continues to drive) rapid change throughout south-east Asia (Chaplot et al., 2002, 2005a, 2005b; Open Development

Mekong, 2019a). Much of this development is uncoordinated, placing the Basin's natural resources – including its water resources – under significant stress (Käkönen, 2008; Johns et al., 2010) and raising concerns about the region's socio-economic and environmental security (Chaplot et al., 2002, 2005a, 2005b; Johns et al., 2010; Barbosa et al., 2016; Hoang et al., 2019).

In 2009, Käkönen and Hirsch reported that, while the Mekong remained one of the world's least-developed major rivers, its perceived 'underused potential' made it a target for a modern development agenda based on large-scale dams and diversions (Käkönen and Hirsch, 2009). Since then, the rate of construction of dams for hydropower generation has markedly accelerated. In 2008, an estimated 2 per cent of mean annual flow was affected; by 2025, the Basin's hydropower reservoir storage is predicted to total over 20 per cent of mean annual flow (Johns et al., 2010). In 2018, more than 120 large hydropower dams were under construction or planned (Dang et al., 2018). While some of these will be 'run-of-river' dams – with limited storage capacity and regulation potential – on the Mekong mainstream (MRC, 2016a), many large-volume reservoirs will be constructed on tributary streams (MRC, 2016a; Brunner et al., 2019). Such impoundments are expected to have major impacts on the river's hydrology (Dang et al., 2018; Hoang et al., 2019), as well as critical fish migration (Ziv et al., 2012) and sediment flows (Schmitt et al., 2017, 2019), with implications for inland capture fisheries and lowland rice production. This will potentially challenge the region's food security and the livelihoods of many people (Barbosa et al., 2016). On the other hand, if well managed with integrated management and timely release of flows (MRC, 2016b), the environmental impacts of reservoirs may be partially mitigated and new opportunities realised for aquaculture and irrigated cropping, resulting in significant socio-economic outcomes, as well as reduced flood risk, for LMB countries (MRC, 2016a).

6.3.3   Agricultural Development and Water Use

One of the sectors likely to benefit from increased water security and redistribution of flows is agriculture (Hall et al., 2014). Agriculture across much of the LMB is dominated by rain-fed rice cultivation and smallholder production systems, with the majority of rice produced also consumed within the region. In recent decades, expansion of large-scale rice cultivation has been occurring in some areas, including the floodplains of northeast Thailand and the Mekong Delta. Both Thailand and Viet Nam now export significant quantities of rice (respectively ranking second and third, globally) as well as a range of other agricultural products (for example, coffee, rubber, sugar, black pepper) to global commodity markets (MRC, 2018a; Australian Trade and Investment

Commission, 2019). Other crops commonly grown in LMB countries include cassava, soybeans, maize, fruits and vegetables (MRC, 2018b).

There is a long history of irrigated farming in the Mekong region, with some traditional systems dating back many hundreds of years (FAO, 2016b). Currently, the proportion of irrigated arable land remains small (for example, in the LMB in 2005/06, this ranged from 9 per cent of cultivated land in Cambodia to 49 per cent in Viet Nam; FAO, 2016b) with some 12 500 irrigation schemes – most of which are designed to provide supplementary irrigation in the wet season – irrigating around 3.6 m ha of farmland (FAO, 2016b). Irrigation is predominantly (98 per cent) from surface water sources (FAO, 2016b). Overall, abstractions for irrigation represent around 12 per cent of annual water supply (Hall et al., 2014); however, access to surface water can be limited in the dry season, when irrigation can represent up to 70 per cent of abstractions (for example, in the Sesan River; Räsänen, 2014) and, in fact, few irrigation systems in the Basin are currently able to operate year-round (FAO, 2016b). In addition, frequent and occasionally devastating floods – exacerbated in recent years by changing rainfall patterns, and unco-ordinated emergency release of water from hydropower reservoirs – have resulted in extensive damage to irrigation infrastructure. For example, in Lao PDR the cost of flood damage to irrigation systems (including lost production) exceeded US$13.8 million in 2018 (GOL, 2018).

In some regions in China, northeast Thailand and the Central Highlands of Viet Nam, groundwater plays a role in mitigating lack of dry season rainfall and surface water deficits, but its use in the Basin is limited (approximately 2 per cent of the total irrigated area) overall (FAO, 2016b). There is considerable scope for expansion in groundwater use, which if well planned and adequately managed could contribute to increased regional and national water (and food) security (e.g., Lacombe et al., 2017); however, this will likely also have impli-cations for hydrological flows in the Mekong (Brunner et al., 2019; Hoang et al., 2019).

6.3.4   Water Quality

There are limited data and specific information, but water pollution in the Mekong is not yet believed to be a serious problem; nevertheless, water contamination, particularly from the agricultural and mining sectors, is an increasing risk to both ecological and human health (Chea et al., 2016). This is particularly true in sub-basins where these activities are concentrated, and for communities who continue to depend on untreated river water for domestic and potable supplies. Government policies are also encouraging a shift from traditional smallholder subsistence farming to modern commercial farming and agrochemical use is consequently increasing, facilitated by the availability of cheap imports of fertilisers and pesticides from China. Research has found pesticide residues (for example, DDT and other organochlorine compounds) in sediment and molluscs in the Viet Nam Delta (Carvalho et al., 2008). In northern Lao PDR, fish-kills have been attributed by local communities and government officials to agro-chemical pollution downstream of irrigated bananas (Radio Free Asia, 2019). In many cases, farmers are unable to read the labelling on Chinese products and so are unaware of the instructions for safe application, with potentially dire consequences for both their own health and the environment.

Finally, mining activities (for example, alluvial gold mining as well as sand and gravel abstraction) are causing adverse environmental and social impacts in places. Concerns relate to mining methods and their impacts on water quality and turbidity, as well as erosion of riverbeds and banks. Communities in southern Laos have complained of toxic water pollution from gold mining in recent years (Radio Free Asia, 2013). As a consequence of increasing environmental and social concerns, in 2012 the government of Lao PDR implemented a nationwide four-year moratorium on new mining projects. Despite pressure from mining companies, the moratorium was extended in 2016 so that investigations into the activities of some concession holders and mining operators could be completed (*Vientiane Times*, 2016).

6.4 DEMAND-SIDE WATER MANAGEMENT AND THE POTENTIAL FOR WATER MARKET DEVELOPMENT

Climate change projections suggest growing water scarcity and variability – thus increasing pressure on water supplies relative to demand – in many regions of the world, including many developing countries. Such countries typically face high levels of exposure to water-related risks, but often have limited ability to cope with or recover from loss and are therefore particularly vulnerable (MacAlister and Subramanyam, 2018). Along with changing climatic conditions, development pressures, including population growth, create novel levels of scarcity (Dore et al., 2010), which is in many respects the case for the Lower Mekong region. In general, where water is plentiful relative to demand, laws regulating water use are likely to be simple and rarely enforced; however, where water is scarce, or increasingly so, more formal institutional systems governing water access and use – that is, processes that effectively regulate access to water resources, determining who gets water, the quality and quantity of water they get and when they get it – will be needed (FAO, 1993; Dore et al., 2010).

Demand-side water policies encompass mechanisms, such as property rights and incentives (for example, penalties, prices), which encourage, and often compel, more socially desirable water use than might otherwise be the case (FAO, 1993; Dore et al., 2010). Historically, command-and-control regulatory systems have prevailed; however, these are increasingly viewed as inflexible and inefficient (Cole and Grossman, 1999; Settre and Wheeler, 2016). The establishment of a market-based water allocation system with tradable water rights (entitlements) is proposed as a cost-effective way to improve the efficiency, equity and sustainability of water use (Settre and Wheeler, 2016). Such water markets require well-defined, enforceable and transferable (that is, tradeable) rights; hence, institutional arrangements must be in place to formalise and secure the water rights held by water users. This provides certainty of water access as well as additional flexibility (enhancing adaptive decision-making and management) and transparency in the system with the opportunity costs of water exposed and many of the negative externalities associated with unregulated water use internalised and reduced (FAO, 1993; Settre and Wheeler, 2016). However, while in developed countries such as Australia it has been shown that decentralised water trading mechanisms have the capacity to move water among users in ways that ensure more efficient allocation of available water supplies and encourage cost-effective water conservation measures (Loch et al., 2013; Mai et al., 2019), it is as yet unclear whether the same applies in all contexts (Rosegrant et al., 2014; Wheeler et al., 2017), or whether this necessarily works seamlessly in practice (Settre and Wheeler, 2016). This is further complicated in transboundary river basins such as the Mekong, which encompass countries with diverse physical, social, cultural, political and economic characteristics.

Using the WMRA framework (Wheeler et al., 2017) as the basis for analysis, we investigated the current state of water resource knowledge, water governance and water management infrastructure in the LMB region. These are recognised as foundational requirements for the establishment of a water market (Matthews, 2004; Wheeler et al., 2017) and constitute essential elements of Stage 1 of the WMRA framework, as detailed earlier in this volume. These elements also represent critical components of sustainable water resource management systems, and hence regional water security (Matthews, 2004; Biggs et al., 2013). While we found no evidence of formal water markets currently operating in the Lower Mekong Basin (hence, Stages 2 and 3 of the WMRA framework are not applicable here), pressure for water policy reform – including adoption of market-based instruments – is being exerted by international development agencies on developing countries in the region. We briefly discuss this and the role of public–private partnerships in progressing more cost-effective water resource management in the region.

6.4.1   Background Information (WMRA Framework Stage 1)

The Mekong River Commission (MRC) is a transboundary inter-governmental river basin organisation established to facilitate joint management and sustainable development of the water resources of the LMB.[1] Formed under the binding 1995 Mekong Agreement, it has a mandate to support water resource planning and negotiations between the four signatory LMB countries (Lao PDR, Cambodia, Thailand and Viet Nam) and, more recently, to promote dialogue on water resource development in the MRB with Myanmar and China.

The MRC works on behalf of its member states to improve water resource planning and governance across the region. Its core responsibility is as an independent scientific knowledge producer, through which it also works to support good governance (Lee and Scurrah, 2009). It acts as an information exchange platform, an investment facilitator and a 'social and environmental guardian of the basin' (Dore and Lazarus, 2009; MRC, 2018a). This overarching role is of particular importance given that its constituent countries are at different points on their development trajectories, and have differing levels of scientific and institutional planning capacity (MRC, 2009a; Carew-Reid, 2017).

6.4.2   Water Knowledge

The MRC plays a central role in water resource management in the Basin through its status as a key knowledge broker (see, e.g., MRC, 2009b, 2009c, 2019b), and its expertise in hydrological modelling and impact assessment of large-scale water resource development projects along the main stem of the Mekong. A decade ago, the 'closed nature' of the MRC's expertise and its role in legitimising a 'modernist agenda' were considered potentially problematic (Käkönen and Hirsch, 2009); more recently, its approach to mainstream development has become more integrated, informed and precautionary (Carew-Reid, 2017). Significantly, the MRC's mandate does not extend to Mekong tributary river systems, where the agendas of individual governments hold sway.

With increasing development pressure on the region's water resources, there has been a broadening of the knowledge base with a number of national and international research organisations also contributing, including the International Water Management Institute (IWMI), part of the inter-governmental Consultative Group on International Agricultural Research (CGIAR) (IWMI, 2019). Satellite technology, in combination with investment in an integrated system of mainstream river height/flow monitoring stations, also provides up-to-date flood monitoring and an improved basis for water resource planning (Hung et al., 2012; MRC, 2019c).

The MRC has developed a number of procedures – including Procedures for the Maintenance of Flows on the Mainstream (PMFM) and Procedures for Water Use Monitoring (PWUM) – designed to support the monitoring of intra-basin water use and inter-basin diversions within the Mekong River system (MRC, 2018a). However, while these were adopted by the MRC Joint Committee in 2006, an integrated monitoring and reporting system is yet to be developed, and varying levels of monitoring and data persist in the Mekong countries. Each of the LMB countries has government ministries, departments and hydrometeorological services responsible for hydrological monitoring and, in some instances, modelling (World Meteorological Organization, 2019). For example, Viet Nam has a National Centre for Hydro-Meteorological Forecasting (NCHMF) under the Ministry of Natural Resources and Environment (MONRE), which collects data from a network of over 230 hydrological stations across 15 major river systems (World Bank and UNISDR, 2013), while Thailand's National Hydroinformatics and Climate Data Center (NHC) in the Ministry of Science and Technology acquires and analyses data from some 1150 major hydrological stations (Vathananukij and Malaikrisanachalee, 2008). However, data access and sharing between LMB countries remains limited amid concerns around national interests and national security (Plengsaeng et al., 2014; Thu and Wehn, 2016), potentially limiting the effective implementation of more integrated management of the region's transboundary water resources (Plengsaeng et al., 2014).

6.4.3   Water Governance

A core focus of the MRC is regional cooperation around the development of the Basin's water resources. As such, with the support of the international community (e.g., Lower Mekong Initiative, n.d.), it works to support good governance within participating countries, enhancing the capacity of local institutions and providing a platform for information exchange, as well as a forum for discussions around more sustainable, equitable and productive water resource management at a range of scales (Hirsch, 2006). This notwithstanding, dialogue around water resources development in the region can be difficult, given the range of different perspectives and competing political interests (Department of Foreign Affairs and Trade, 2015). The MRC is currently developing and implementing an integrated water resources management (IWRM) programme, funded through the World Bank (2019b), aimed at addressing a number of critical transboundary issues, including coordination of flows to better meet the requirements of downstream users and environments. At a broader sub-continental scale, MRB countries also participate in the Water Environment Partnership in Asia (WEPA), which provides further opportunity for information and knowledge-sharing on water environmental governance issues (Ministry of the Environment of Japan, 2019).

At the national level, each of the five LMB countries has multiple government departments, policies and laws dealing with water resource management, but all face challenges in implementing these and, excepting Myanmar, ensuring that decision-making conforms to the requirements of the Mekong Agreement, particularly with accelerating pressure to develop the Basin's natural resources (Öjendal and Jensen, 2012). Key governance challenges variously include: inadequate legislative frameworks; limited coordination (including data-sharing) among water-related institutions at both national and regional scales; weak water resource planning and implementation; unplanned urban and industrial development; and limited safeguards/inadequate protection for local communities, agricultural production systems and natural ecosystems (Suhardiman et al., 2012; FAO, 2016a; Broegaard et al., 2017).

Poor governance has significant implications for social and environmental equity, particularly where governments also have a short-term economic development agenda (Matthews, 2012). This is compounded where property rights are either non-existent or poorly documented (Neef et al., 2006). While Grindle (2004) argues the case for 'good enough' governance in developing countries, others point to the ongoing risk to national economies and the livelihoods of local people (Sithirith, 2017). Such issues are likely to constrain the effective operation of efficient resource allocation mechanisms such as water markets (Wheeler et al., 2017) and schemes for payments for ecosystem services (Arias et al., 2011).

Explicit, hence tradeable, property (land and water) rights are an important foundation for market-based systems, which can, when well designed such that free market forces are allowed to operate and in the case of water resources, ensure a more efficient, equitable and secure water future (Matthews, 2004). The geo-politics of the LMB region – with single-party states operating in Cambodia, Lao PDR and Viet Nam and a strong military influence on government in Myanmar and Thailand (Freedom House, 2019) – will likely constrain the effective operation of such market-based instruments. Nevertheless, some improvements are in evidence. For example, in Viet Nam, Lao PDR and Myanmar, land is largely state-owned and subject to a high level of central government control, but landholders have customary usage rights (Jepsen et al., 2019); in Viet Nam, these rights are now tradeable, and hence approximate land ownership (Neef et al., 2006; Marsh et al., 2007), offering some hope for future change.

6.4.4   Infrastructure

In terms of key water resource management infrastructure, the Consultative Group on International Agricultural Research (CGIAR)[2] maintains a database of dams (hydropower dams ≥15 megawatts installed capacity, and irrigation reservoirs ≥ 0.5 km[2]) for the entire Mekong River Basin (CGIAR, 2019), while the MRC hosts an irrigation database for the LMB based on national-level information sourced from its constituent countries (MRC, 2018b). Currently, the MRC irrigation database identifies a total of 6755 irrigation projects[3] in the Lower Mekong Basin (excluding Myanmar), servicing a currently irrigable area of 4 780 730 ha of (largely) rice crops. Infrastructure includes 6596 existing irrigation headworks; 1317 irrigation reservoirs with an irrigation capacity of over 11 378 million m[3],[4] and an extensive network of canals in the Mekong Delta (MRC, 2018b).

Since the 1950s, modernisation of irrigation systems has occurred in most LMB countries, aimed at reducing vulnerability to drought and increasing crop yields and national food security and self-sufficiency (FAO, 2016b). That said, geopolitical conflict in the region through the 1960s, 1970s and 1980s stymied progress, and many systems have fallen into disrepair due to limited resources (e.g., Sithirith, 2017). All LMB countries have current policies and plans for expansion of irrigation, but all also, to varying degrees, face a range of financial, technological and institutional/organisational challenges associated with the upgrading and maintenance of irrigation infrastructure and equipment, and the regulation of access to and use of water resources (Hall et al., 2014; FAO, 2016b; Sithirith, 2017; MRC, 2018b).

International finance[5] plays a vital role in facilitating improvements in water management (and other) infrastructure in the region (Open Development Mekong, 2019b). In particular, investments in major construction projects directed at the agricultural and energy sectors are likely to result in increased surface water storage capacity in the system, and the ability to redistribute a portion of wet season river flows to the dry season (MRC, 2009a). While such finance is generally framed in the context of addressing poverty and other Sustainable Development Goals (SDGs) through agricultural and economic expansion, it also has the capacity to drive progress in terms of institutional, policy and regulatory development (MRC, 2016a; United Nations, 2017). However, the increase in opportunistic private sector participation also presents a significant challenge in the region, where effective regulatory mechanisms are not yet in place to ensure that development results in sustainable socio-economic and environmental outcomes (MRC, 2016a; Sithirith, 2017).

In a review of water governance in Cambodia, Sithirith (2017) discusses the state of irrigation infrastructure, concluding that investment in large-scale water projects over recent decades has often failed to deliver the promised level of economic growth, due to their misalignment with the needs of local people and the lack of ongoing funding for their operation and maintenance, resulting in a significant proportion of systems falling into disrepair. He argues the case for smaller-scale infrastructure investments, developed in conjunction with and managed by local user groups, as a way forward that reduces conflict, improves the equitable sharing of water resources and results in greater water productivity (Sithirith, 2017). Similarly, in Lao PDR, many irrigation schemes are either under-performing due to poor water management or deteriorating due to poor maintenance and/or frequent flood damage. In addition, while fees may be collected for irrigation services, these are often insufficient to cover operating costs, despite government subsidies, resulting in substantial accumulated debt (IWMI and WorldFish, 2019).

6.4.5   Water Resource Management: The Way Forward

Water underpins economic growth and development in mainland southeast Asia, including countries in the Mekong region. However, as discussed above, water governance to ensure sustainable, equitable and cost-effective access and use of available water resources is a key challenge, both nationally and regionally (Hirsch, 2006; Dore et al., 2010). While elsewhere (for example, Australia) water markets operate to enable redistribution of water between buyers and sellers within a connected hydrological system in a bid to ensure its highest-value use, there is currently no evidence of formal water markets in the Mekong Region (Dore et al., 2010; ADB, 2017). This is despite both the World Bank (WB) and the Asian Development Bank (ADB), which help to finance many large infrastructure investments in the region, increasingly using their investments to leverage policy change towards increased or full cost recovery through 'economic' pricing, privatisation of public sector institutions and services, and the operation of free market mechanisms such as water markets to ensure priority of access to sectors or user groups able to demonstrate high returns from water (Siregar, 2004).

Long involved in the Asian water sector, the ADB promotes water as a socially vital economic good underpinning equitable economic growth and poverty reduction; however, moves to make ADB loans contingent on water policy reforms which adopt free market approaches may, in effect, work to undermine these goals (Siregar, 2004). The privatisation of services, along with user fees charged to cover the costs of operation, maintenance and capital expenditure, can result in reduced access to good-quality water resources for poor and marginalised groups in society, including many smallholder farmers. The implementation of cost recovery and market mechanisms for water allocation may also thereby endanger food sovereignty in these regions. For instance, paddy rice cultivation (a main source of domestic food) requires relatively high levels of water consumption and returns less revenue than other cash crops or industrial uses; reallocation of water to where it delivers the greatest economic advantage will likely threaten paddy cultivation, hence also regional food security (Siregar, 2004).

On the other hand, the ADB makes a case for the privatisation (or at least private–public partnerships) of irrigation infrastructure. The declining condition of large surface irrigation schemes (discussed above), with system failures and deteriorating infrastructure due to limited operation and maintenance budgets, reduces the reliability of water supply, hence the capacity and willingness of farmers to plant more valuable crops (ADB, 2017). The ADB suggests that the problem is due, in part, to the high administration costs of the irrigation bureaucracy ('a powerful force in maintaining the status quo'), limiting the funds made available for ongoing maintenance (ADB, 2017). A reformed irrigation system with greater security (reliability and flexibility) of supply – particularly one cognisant of farmers' needs (that is, fit for purpose) – might see farmers more able and willing to pay more for their water; adequate funding for operation and maintenance of the system; even better services and systems; and have the potential to attract additional sources of finance for technology and practice change (ADB, 2017). In instances where effective stakeholder-informed private–public partnerships can be developed, there may also be scope for localised informal water trading systems (Bjornlund, 2004), particularly where higher-value crops can be grown to achieve both higher water productivity and increased adaptive capacity; however, in the absence of secure and tradable water rights (as discussed earlier), it is unlikely that these can operate in more formalised ways that require effective institutional and administrative arrangements (Bjornlund, 2004).

6.5 CONCLUSIONS

With a significant development agenda in play in the transboundary Lower Mekong River Basin of mainland southeast Asia, there is increasing pressure on regional water resources and on LMB countries – with the aid of the MRC – to ensure their equitable and sustainable management. The WMRA framework (Wheeler et al., 2017) provides a useful basis for assessing whether market-based instruments such as water markets might play a role – as they have done in other jurisdictions – specifically in demand-side management and the efficient allocation and use of water in the region. Strong knowledge networks, supporting integrated water resource management, along with robust governance and well-planned and financed functional water management infrastructure, which meets the needs of local producers/user groups, are needed to avoid the risk of over-allocation and degradation of the resource, and of detrimental ecological and socio-economic impacts, especially amongst smallholder communities.

We reviewed the state of knowledge of hydrological systems, water resources governance and infrastructure in the LMB countries; as discussed above, these are considered essential foundations (that is, Stage 1) for effective water markets. While progress on building these foundations is apparent, there is also evidence of significant structural shortcomings.

Currently, the social, political and economic diversity of the countries means that despite the presence of the MRC, LMB governments are yet to fully embrace an integrated approach to the management of water resources. From a basin perspective, the result is suboptimal decision-making, with adverse implications for economic returns and also the resilience and sustainability of investments, and the livelihoods and well-being of those directly impacted. Typically, water resource interventions fall short in protecting key environmental assets and ecosystem services and, frequently, the benefits, impacts and risks from development interventions are inequitably distributed. Inability to adequately manage floods and droughts continues to adversely impact livelihoods and undermines socio-economic development. In the context of the LMB, it seems unlikely that market-based approaches will be practicable or successful, at least in the near future. Instead, more conventional regulatory approaches, along with moves currently afoot for integrated basin planning and management, will likely better serve the communities and environments of the region.

ACKNOWLEDGEMENTS

This chapter was produced with the support of the German Federal Ministry for the Environment, Nature Conservation, Building and Nuclear Safety through the International Climate Initiative, the University of Southern Queensland Centre for Applied Climate Sciences and the CGIAR programme Water, Land and Ecosystems. We thank members of the Mekong River Commission (MRC) Secretariat in Vientiane, Lao PDR, for valuable discussions and permission to reproduce the map in Figure 6.1 and, along with other colleagues – in particular, Dr Adam Loch and Professor Shahbaz Mushtaq – for advice on the manuscript.

NOTES

1. Geographically, the Lower Mekong Basin (LMB) includes Myanmar; however, Myanmar is not one of the four LMB countries that are signatories to the 1995 Mekong Agreement which is enacted through Mekong River Commission (MRC).

2. The CGIAR is a global partnership of international organizations engaged in food security research to address poverty, hunger and environmental degradation. https://www.cgiar.org/.
3. This value excludes small and medium-scale projects in Thailand (MRC, 2018b).
4. This value excludes Cambodia, for which no reservoir capacity information is available (MRC, 2018b).
5. Sources include concessional loans from multilateral development institutions such as the Asian Development Bank and World Bank; foreign aid or grants from countries such as Japan, the United States, China, France and Australia; and other forms of financing through public–private partnerships and privately financed build–own–transfer models.

REFERENCES

Adamson, P., and Bird, J. (2010). The Mekong: a drought-prone tropical environment? *International Journal of Water Resources Development, 26*(4), 579–594.
Arias, M.E., Cochrane, T.A., Lawrence, K.S., Killeen, T.J., and Farrell, T.A. (2011). Paying the forest for electricity: a modelling framework to market forest conservation as payment for ecosystem services benefiting hydropower generation. *Environmental Conservation, 38*(4), 473–484.
Asian Development Bank (ADB) (2017). *Financing Asian Irrigation: Choices before Us.* Asian Development Bank. https://www.adb.org/sites/default/files/publication/306856/financing-asian-irrigation.pdf.
Australian Trade and Investment Commission (2019). *Export Markets – Vietnam.* Australian Trade and Investment Commission. https://www.austrade.gov.au/australian/export/export-markets/countries/vietnam/industries/agribusiness.
Barbosa, C.C.D., Dearing, J., Szabo, S., Hossain, S., Binh, N.T., et al. (2016). Evolutionary social and biogeophysical changes in the Amazon, Ganges–Brahmaputra–Meghna and Mekong deltas. *Sustainability Science, 11*(4), 555–574.
Biggs, E.M., Duncan, J.M.A., Atkinson, P.M., and Dash, J. (2013). Plenty of water, not enough strategy: how inadequate accessibility, poor governance and a volatile government can tip the balance against ensuring water security: the case of Nepal. *Environmental Science and Policy, 33*, 388–394.
Bjornlund, H. (2004). Formal and informal water markets: drivers of sustainable rural communities? *Water Resources Research, 40*(9), 1–12.
Broegaard, R.B., Vongvisouk, T., and Mertz, O. (2017). Contradictory land use plans and policies in Laos: tenure security and the threat of exclusion. *World Development, 89*, 170–183.
Brunner, J., Carew-Reid, J., Glemet, R., McCartney, M., and Riddell, P. (2019). *Measuring, Understanding and Adapting to Nexus Trade-Offs in the Sekong, Sesan and Srepok Transboundary River Basins.* IUCN, Viet Nam Country Office, 82 pp.
Carew-Reid, J. (2017). The Mekong: strategic environmental assessment of mainstream hydropower development in an international river basin. In P. Hirsch (ed.), *Handbook of the Environment in Southeast Asia* (pp. 334–355). Routledge.
Carvalho, F., Villeneuve, J.P., Cattini, C., Tolsa, I., Thuan, D.D., and Nhan, D.D. (2008). Agrochemical and polychlorobyphenyl (PCB) residues in the Mekong River delta, Vietnam. *Marine Pollution Bulletin, 56*, 1476–1485.

Chaplot, V., Boonsaner, A., Bricquet, J.P., De Rouw, A., Janeau, J.L., et al. (2002). Soil erosion under land use change from three catchments in Laos, Thailand and Vietnam. *12th ISCO Conference, Beijing* (pp. 313–318).

Chaplot, V., Coadou le Brozec, E., Silvera, N., and Valentin, C. (2005a). Spatial and temporal assessment of linear erosion in catchments under sloping lands of northern Laos. *Catena, 63*(2), 167–184.

Chaplot, V., Giboire, G., Marchand, P., and Valentin, C. (2005b). Dynamic modelling for linear erosion initiation and development under climate and land-use changes in northern Laos. *Catena, 63*(2), 318–328.

Chea, R., Grenouillet, G., and Lek, S. (2016). Evidence of water quality degradation in Lower Mekong Basin revealed by self-organizing map. *PLos ONE, 11*(1), e0145527–e0145527.

Cole, D.H., and Grossman, P.Z. (1999). When is command-and-control efficient? Institutions, technology, and the comparative efficiency of alternative regulatory regimes for environmental protection. Articles by Maurer Faculty, Paper 590, Maurer School of Law, Indiana University.

Consultative Group on International Agricultural Research (CGIAR) (2019). *Greater Mekong Dams Observatory.* CGIAR Research Program on Water, Land and Ecosystems. http://wle-mekong.cgiar.org/changes/our-research/greater-mekong -dams-observatory/.

Cook, B.I., Bell, A.R., Anchukaitis, K.J., and Buckley, B.M. (2012). Snow cover and precipitation impacts on dry season streamflow in the Lower Mekong Basin. *Journal of Geophysical Research: Atmospheres, 117*(D16), 1–11.

Dang, T.D., Cochrane, T.A., Arias, M.E., and Tri, V.P.D. (2018). Future hydrological alterations in the Mekong Delta under the impact of water resources development, land subsidence and sea level rise. *Journal of Hydrology – Regional Studies, 15*, 119–133.

Department of Foreign Affairs and Trade (2015). *Australia's Mekong Water Resources Program.* Australian Department of Foreign Affairs and Trade (DFAT), 5 pp.

Dore, J., and Lazarus, K. (2009). De-marginalizing the Mekong River Commission. In F. Molle, T. Foran and M. Käkönen (eds), *Contested Waterscapes in the Mekong Region: Hydropower, Livelihoods and Governance* (pp. 357–381). Earthscan.

Dore, J., Molle, F., Lebel, L., Foran, T., and Lazarus K. (eds) (2010). *Improving Mekong Water Resources Investment and Allocation Choices.* Project Report (Project Number PN 67), CGIAR Challenge Program on Water and Food (CPWF).

Eastham, J., Mpelasoka, F., Mainuddin, M., Ticehurst, C., Dyce, P., et al. (2008). *Mekong River Basin Water Resources Assessment: Impacts of Climate Change.* Water for a Healthy Country Flagship Report. Commonwealth Scientific and Industrial Research Organisation (CSIRO), 153 pp.

Food and Agriculture Organization (FAO) (1993). *The State of Food and Agriculture.* Food and Agriculture Organization of the United Nations, 377 pp.

Food and Agriculture Organization (FAO) (2011). *Irrigation in Southern and Eastern Asia in Figures AQUASTAT Survey – 2011.* FAO Water Report 37, United Nations Food and Agricultural Organization of the United Nations, 512 pp.

Food and Agriculture Organization (FAO) (2016a). *Institutional Framework.* Food and Agriculture Organization of the United Nations. http://www.fao.org/nr/water/ aquastat/watermpl/index.stm.

Food and Agriculture Organization (FAO) (2016b). *Irrigation and Drainage.* Food and Agriculture Organization of the United Nations. http://www.fao.org/nr/water/ aquastat/watermpl/index.stm.

Freedom House (2019). *Freedom in the World 2019*. Democracy in Retreat, Freedom House. https://freedomhouse.org/report/freedom-world/freedom-world-2019/democracy-in -retreat.

Ge, F., Zhi, X., Babar, Z.A., Tang, W., and Chen, P. (2017). Interannual variability of summer monsoon precipitation over the Indochina Peninsula in association with ENSO. *Theoretical and Applied Climatology*, *128*(3), 523–531.

Gebretsadik, Y., Fant, C., and Strzepek, K. (2012). Impact of climate change on irrigation, crops and hydropower in Vietnam. WIDER Working Paper 2012/79, 30 pp.

Government of Lao PDR (GOL) (2018). *Flood Post Disaster Needs Assessment*. Final Report, Government of Lao PDR, 158 pp.

Grainger, C., Köhlin, G., Coria, J., Whittington, D., Xu, J., et al. (2019). *Scaling Up Private Sector Participation and Use of Market-Based Approaches for Environmental Management: Opportunities for Scaling Up Market-Based Approaches to Environmental Management in Asia*. Project 49354-001, Report prepared by the Environment Thematic Group, Sustainable Development and Climate Change Department. Asian Development Bank (ADB), 301 pp.

Grindle, M.S. (2004). Good enough governance: poverty reduction and reform in developing countries. *Governance: An International Journal of Policy, Administration, and Institutions*, *17*(4), 525–548.

Hall, B., Minami, I., Jantakad, P., and Dinh, C.N. (2014). *Irrigation for Food Security, Poverty Alleviation and Rural Development in the LMB*. Vientiane, Lao PDR, Mekong River Commission (MRC), 54 pp.

Harris, L.M., Rodina, L., and Morinville, C. (2015). Revisiting the human right to water from an environmental justice lens. *Politics, Groups, and Identities*, *3*(4), 660–665.

Hecht, J.S., Lacombe, G., Arias, M.E., Dang, T.D., and Piman, T. (2019). Hydropower dams of the Mekong River basin: a review of their hydrological impacts. *Journal of Hydrology*, *568*, 285–300.

Hirsch, P. (2006). Water governance reform and catchment management in the Mekong Region. *Journal of Environment and Development*, *15*(2), 184–201.

Hoang, L.P., van Vliet, M.T.H., Kummu, M., Lauri, H., Koponen, J., et al. (2019). The Mekong's future flows under multiple drivers: how climate change, hydropower developments and irrigation expansions drive hydrological changes. *Science of the Total Environment*, *649*, 601–609.

Hoanh, C.T., Facon, T., Thuon, T., Bastakoti, R.C., Molle, F., and Phengphaengsy, F. (2009). Irrigation in the Lower Mekong Basin countries: the beginning of a new era? In F. Molle, T. Foran and M. Käkönen (eds), *Contested Waterscapes in the Mekong Region: Hydropower, Livelihoods and Governance* (pp. 143–171). Earthscan.

Hung, N.N., Delgado, J.M., Tri, V.K., Hung, L.M., Merz, B., et al. (2012). Floodplain hydrology of the Mekong Delta, Vietnam. *Hydrological Processes*, *26*(5), 674–686.

International Water Management Institute (IWMI) (2019). What we do, Colombo, Sri Lanka: International Water Management Institute. http://www.iwmi.cgiar.org/.

IWMI and WorldFish (2019). Lao PDR Irrigation Subsector Review. ADB Grant No. 0534-LAO Project No. 42203-025.

Jepsen, M.R., Palm, M., and Bruun, T.B. (2019). What awaits Myanmar's uplands farmers? Lessons learned from mainland south-east Asia. *Land*, *8*(2), 29.

Johns, F., Saul, B., Hirsch, P., Stephens, T., and Boer, B. (2010). Law and the Mekong River Basin: a socio-legal research agenda on the role of hard and soft law in regulating transboundary water resources. *Melbourne Journal of International Law*, *11*, 1–21.

Käkönen, M. (2008). Mekong Delta at the crossroads: more control or adaptation? *Ambio, 37*(3), 205–212.

Käkönen, M., and Hirsch, P. (2009). *The Anti-Politics of Mekong Knowledge Production.* Earthscan.

Lacombe, G., Douangsavanh, S., Vongphachanh, S., and Pavelic, P. (2017). Regional assessment of groundwater recharge in the Lower Mekong Basin. *Hydrology, 4*(4), 60.

Lee, G., and Scurrah, N. (2009). *Power and Responsibility: The Mekong River Commission and Lower Mekong Mainstream Dams.* University of Sydney.

Loch, A., Wheeler, S.A., Bjornlund, H., Beecham, S., Edwards, J., et al. (2013). *The Role of Water Markets in Climate Change Adaptation.* National Climate Change Adaptation Research Facility (NCCARF), 125 pp.

Lower Mekong Initiative (n.d.). *Lower Mekong Initiative: Friends of the Lower Mekong.* https://www.lowermekong.org/partner/background-and-approach.

MacAlister, C., and Subramanyam, N. (2018). Climate change and adaptive water management: innovative solutions from the global South. *Water International, 43*(2), 133–144.

Mai, T., Mushtaq, S., Loch, A., Reardon-Smith, K., and An-Vo, D.-A. (2019). A systems thinking approach to water trade: finding leverage for sustainable development. *Land Use Policy, 82*, 595–608.

Marsh, S.P., MacAulay, T.G., and Hung, P.V. (eds) (2007). *Agricultural Development and Land Policy in Vietnam: Policy Briefs.* ACIAR Monograph No. 126, Australian Centre for International Agricultural Research (ACIAR), 72 pp.

Matthews, N. (2012). Water grabbing in the Mekong Basin: an analysis of the winners and losers of Thailand's hydropower development in Lao PDR. *Water Alternatives, 5*(2), 392–411.

Matthews, O. (2004). Fundamental questions about water rights and market reallocation. *Water Resources Research, 40*, W09S08.

Mekong River Commission (MRC) (2009a). *Adaptation to Climate Change in the Countries of the Lower Mekong Basin.* MRC Management Information Series No. 1, Mekong River Commission, 8 pp.

Mekong River Commission (MRC) (2009b). *The Flow of the Mekong.* MRC Management Information Series No. 2, Mekong River Commission, 12 pp.

Mekong River Commission (MRC) (2009c). *Modelling the Flow of the Mekong.* MRC Management Information Series No. 3, Mekong River Commission, 12 pp.

Mekong River Commission (MRC) (2016a). *Integrated Water Resources Management-based Basin Development Strategy 2016–2020 for the Lower Mekong Basin.* Mekong River Commission, 108 pp.

Mekong River Commission (MRC) (2016b). *Development of Guidelines for Hydropower Environmental Impact Mitigation and Risk Management in the Lower Mekong Mainstream and Tributaries.* The ISH 0306 Study, Mekong River Commission, 261 pp.

Mekong River Commission (MRC) (2018a). *An Introduction to MRC Procedural Rules for Mekong Water Cooperation.* Mekong River Commission, 20 pp.

Mekong River Commission (MRC) (2018b). *Irrigation Database Improvement for the Lower Mekong Basin.* MRC Technical Report No. 1, Mekong River Commission, 122 pp.

Mekong River Commission (MRC) (2019a). Mekong water levels reach low record. Mekong River Commission. http://www.mrcmekong.org/news-and-events/news/mekong-water-levels-reach-low-record/.

Mekong River Commission (MRC) (2019b). *Data and Information Services.* Mekong River Commission. http://portal.mrcmekong.org/index.
Mekong River Commission (MRC) (2019c). *Mekong Flood Forecasting.* Mekong River Commission. http://www.mrcmekong.org/.
Ministry of the Environment of Japan (2019). *Water Environment Partnership in Asia.* http://wepa-db.net/3rd/en/about.html.
National Geographic (2019). Mekong River at its lowest in 100 years, threatening food supply. https://www.nationalgeographic.com/environment/2019/07/mekong-river -lowest-levels-100-years-food-shortages/.
Neef, A., Hager, J., Wirth, T., Schwarzmeier, R., and Heidhues, F. (2006). Land tenure and water rights in Thailand and Vietnam: challenges for ethnic minorities in moun- tainous forest regions. *Geographica Helvetica, 61*(4), 255–265.
Öjendal, J., and Jensen, K.M. (2012). Politics and development of the Mekong River Basin: transboundary dilemmas and participatory ambitions. In J. Öjendal, S. Hansson and S. Hellberg (eds), *Politics and Development in a Transboundary Watershed* (pp. 37–59). Springer.
Open Development Mekong (2019a). *Population and Censuses.* Open Development Initiative, East-West Management Institute (EWMI).
Open Development Mekong (2019b). *Infrastructure, Open Development Initiative.* East–West Management Institute (EWMI).
Plengsaeng, B., When, U., and van der Zaag, P. (2014). Data-sharing bottlenecks in transboundary integrated water resources management: a case study of the Mekong River Commission's procedures for data sharing in the Thai context. *Water International, 39*(7), 933–951.
Radio Free Asia (2013). Lao dams, mining ruining Mekong water quality in Cambodia. http://www.rfa.org/english/news/laos/sekong-06252013190742.html.
Radio Free Asia (2019). Chinese banana plantations in Lao District leave locals with little land to farm. https://www.rfa.org/english/news/laos/plantations -03052019144331.html.
Räsänen, T.A. (2014). Hydrological changes in the Mekong River Basin: the effects of climate variability and hydropower development. PhD Thesis, Aalto University, Finland, 69 pp.
Richter, B. (2016). Water markets can support an improved water future. The Water Blog, World Bank. https://blogs.worldbank.org/water/water-markets-can-support -improved-water-future.
Rosegrant, M.W., Ringler, C., and Zhu, T. (2014). Water markets as an adaptive response to climate change. In K.W. Easter and Q. Huang (eds), *Water Markets for the 21st Century: What Have We Learned?* (pp. 35–55). Springer.
Schmitt, R.J.P., Kittner, N., Kondolf, G.M., and Kammen, D.M. (2019). Deploy diverse renewables to save tropical rivers. *Nature, 569,* 330–332.
Schmitt, R.J.P., Rubin, Z., and Kondolf, G.M. (2017). Losing ground – scenarios of land loss as consequence of shifting sediment budgets in the Mekong Delta. *Geomorphology, 294,* 58–69.
Settre, C., and Wheeler, S.A. (2016). Environmental water governance in the Murray– Darling Basin of Australia. In V. Ramiah and G.N. Gregoriou (eds), *Handbook of Environmental and Sustainable Finance* (pp. 67–91). Elsevier.
Siregar, P.R. (2004). World Bank and ADB's role in privatising water in Asia. Committee for the Abolition of Illegitimate Debt (CADTM). http://www.cadtm.org/ spip.php?page=imprimer&id_article=544.

Sithirith, M. (2017). Water governance in Cambodia: from centralised water governance to farmer water user community. *Resources*, *6*, 44.
Suhardiman, D., Giordano, M., and Molle, F. (2012). Scalar disconnect: the logic of transboundary water governance in the Mekong. *Society and Natural Resources*, *25*(6), 572–586.
Thu, H.N., and Wehn, U. (2016). Data sharing in international transboundary contexts: the Vietnamese perspective on data sharing in the Lower Mekong Basin. *Journal of Hydrology*, *536*, 351–364.
United Nations (2017). *Financing for Development: Progress and Prospects*. United Nations (UN) Inter-agency Task Force on Financing for Development, 139 pp.
United Nations University (2013). *Water Security and the Global Water Agenda*. A UN-Water Analytical Brief, United Nations University Institute for Water, Environment and Health (UNU-INWEH), 47 pp.
Vathananukij, H., and Malaikrisanachalee, S. (2008). Hydroinformatic system (implementation in Thailand). *Water SA*, *34*(6), 725–730.
Vientiane Times (2016). PM announces continued suspension of mining concessions. http://www.vientianetimes.org.la/FreeContent/FreeConten_PM245.htm.
Villafuerte, M.Q., and Matsumoto, J. (2015). Significant influences of global mean temperature and ENSO on extreme rainfall in south-east Asia. *Journal of Climate*, *28*(5), 1905–1919.
Wheeler, S.A., Loch, A., Crase, L., Young, M., and Grafton, R.Q. (2017). Developing a water market readiness assessment framework. *Journal of Hydrology*, *552*, 807–820.
World Bank (2019a). *Water*. https://www.worldbank.org/en/topic/water/overview#1.
World Bank (2019b). *Mekong Integrated Water Resources Management*. http://projects.worldbank.org/P104806/mekong-integrated-water-resources-management?lang=en&tab=overview.
World Bank and UNISDR (2013). *Strengthening of Hydrometeorological Services in Southeast Asia*. Country Assessment Report for Viet Nam, 101 pp.
World Meteorological Organization (2019). *Country Profile Database*. World Meteorological Organization (WMO). https://cpdb.wmo.int/.
Ziv, G., Baran, E., Nam, S., Rodriguez-Iturbe, I., and Levin, S.A. (2012). Trading-off fish biodiversity, food security, and hydropower in the Mekong River Basin. *PNAS*, *109*(15), 5609–5614.

7. Nepal: a country where water policy is in flux

Andrew Johnson, Madhav Belbase, Keshab Dhoj Adhikari, Maheswor Shrestha and Juliane Haensch

7.1 INTRODUCTION

Nepal is considered a water-surplus country in the Hindu Kush–Himalaya (HKH) region which is the source of water for over 1 billion people, or more than 15 per cent of the world's population, and is the source of ten of the world's major river systems (Wester et al., 2019). However, Wester et al. (2019) clearly indicate that the effects of climate change on the region including Nepal will be dramatic. The report notes that a 1.5°C rise in temperature (already observed in the region) will lead to a 30 per cent reduction in snow and ice reserves, while this will be made up in the short term by increased monsoonal rainfall. The pattern of precipitation is also predicted to change significantly from a regular well-predicted pattern to a much more erratic pattern with peaks and troughs. The consequences will be periods of intense rainfall and floods followed by relatively dry periods. In this region, it is clear, assessments of water availability based on historic records will not necessarily be good predictors of water availability in the future. Figure 7.1 shows the importance of the HKH region as the source of some of the most important river systems in the region, especially the Indus, Ganges and Brahmaputra, as well as the Amu Darya, Yangtze, Mekong and Irrawaddy.

Nepal is a country of about 28 million people, which is dominated by the capital Kathmandu (World Bank, 2019). Its landscape is dominated by high mountain peaks, including eight of the ten highest mountains in the world (for example, Mount Everest – Sagarmartha) but includes lowland plains (Terai). Consequently, it includes agro-ecological zones capable of producing an amazing variety of agricultural and horticultural crops. The Terai is particularly productive and is a focus of a range of irrigation areas as well as canals providing considerable water for irrigation systems in India. Employment in

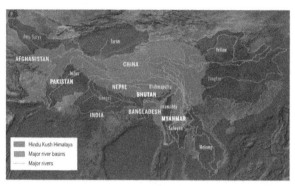

Source: Wester et al. (2019).

Figure 7.1   *The Hindu Kush-Himalayan region and ten major river basins*

agriculture reaches 70 per cent (of total employment); 29 per cent of Nepal's land is in agriculture; and the sector of agriculture, forestry and fishing adds 25 per cent to Nepal's gross domestic product (GDP) (World Bank, 2019). In the Nepali context, water is considered a critical economic driver for hydropower generation, irrigation, water supply and sanitation, and water-induced disaster management (e.g., OECD, 2003).

Nepal has abundant river resources. The four main river systems are the Mahakali (shared with India), Karnali, Gandaki and Kosi (connected with China), which originate in the Himalayas and have significant discharge even in the dry season. Medium river basins include the Babai, West Rapti, Bagmati, Kamala, Kankai and the Mechi, which are characterized by wide seasonal fluctuations. Furthermore, there are a few minor rivers and tributaries (FAO, 2012; WECS, 2011). Nepal's vast water resources offer substantial hydropower and irrigation potential. However, Nepal uses only a small proportion of its hydropower potential, has built few surface water storages, and irrigates only 70 per cent of its irrigable land (ADB, 2013). Hence, Nepal's water sector holds an important development potential in the continuing policy planning discourse (Pakhtigian et al., 2019).

Nepal is one of the least-developed countries in the world and has in recent years been the focus of considerable political change (see Table 7.1) which, it could be argued, has limited the development of water resources in the country. The current period, in which the country seems to lurch towards a lasting political stability, has necessitated a relatively rapid process of water sector reform owing mainly to the state restructuring following the promulgation of the new

*Table 7.1   Nepal: recent political history of government in Nepal
             (monarchy to democratic elections)*

1990	Monarch agrees to restore multi-party democracy
1996	Maoist insurgency commenced following political instability
2001	Royal family massacre: King, Queen and other royals killed
2006	Comprehensive peace agreement
2007	Interim government established and monarchy abolished
2008	Nepal becomes a republic
2015 April	Major earthquake
2015 Sept	New Constitution adopted creating three levels of government (previously only national)
2016 Dec	Draft national Water Resources Policy released for public consultation
2017 May	Election of new local government representatives
2017 Nov/Dec	Election of national and provincial government representatives
2018 Jan	New provincial assemblies established
2018 Feb	New national government formed (communist majority)

Constitution in 2015. The new Constitution contains a specific policy related to the use of natural resources including water. It also contains a range of complementary provisions which influence the management of water resources within the country (NLC, 2018a).

Historically, water rights were not regulated in detail, as water was not considered an important resource and water rights were linked with land rights. Between 1961 and 1992, water resources gained importance in the Nepalese political economy and their ownership was vested with the state (Shrestha, 2009).

The Constitution secured the fundamental right for every citizen to live in a clean and healthy environment as well as to have access to drinking water and sanitation. Similarly, while considering the values conducive to national welfare and inter-generational equity, the formulation of the constitutional policy relating to natural resources clearly outlines a focus on the protection, promotion and environmentally sustainable use of available natural resources. While the Constitution establishes a framework for the management of natural resources, the delivery of these functions is not necessarily clear at this stage, given that it has provided for the sharing of power and responsibility among the national, provincial and local levels of government (see Figure 7.2, which demonstrates the seven provinces).

In recent years, as shown in Table 7.1, the country has endorsed a new Constitution as well as undertaken elections at all three levels. The elections resulted in the establishment of elected governments at all levels. Likewise, administrative functions of the bureaucracy have also recently been established

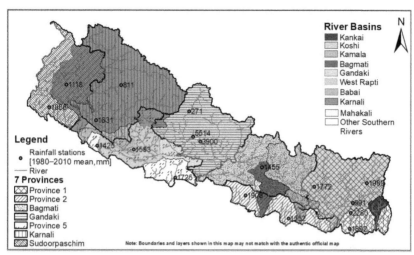

Source: WECS and CSIRO (2019).

Figure 7.2   *Nepal's major river basins, seven provinces and spatial distribution of point rainfall stations*

at all three levels; note that each has some responsibility for water management (in some situations concurrent national/provincial powers are anticipated).

Simplistically, the breakdown of responsibilities as outlined in the Constitution are: national: responsible for major policy and inter-jurisdictional measures; national/provincial: concurrent powers for certain activities such as irrigation and water supply; provincial: responsible for operational management; and municipal: responsible for delivery at the local level.

Organizationally, it is anticipated that a national Water Resources Policy will outline the goals and principles for water resources management across the country. New legislation should give effect to the policy, enable its implementation and define the relationship between the three levels of government. It is anticipated that subsequent provincial water resources should also give effect to the national policy in a more focused manner and enable delivery at the provincial level. Based on experience in Australia, it is considered that in a federated system it is important for the central and provincial governments to develop consistent approaches to water resources management (Wheeler, 2014). Consultation has commenced between the centre and provincial water managers to identify an agreed way forward.

The water markets readiness assessment (WMRA) framework (Wheeler et al., 2017) identifies conditions necessary to facilitate water trading for regions.

With the help of this framework, potential barriers to the implementation of water markets can be identified. At this stage, Nepal's government is not planning to introduce formal water markets. On the assessment of its institutional conditions, Nepal is situated in step one (1a) of the WMRA framework, concerning the calculation/definition of the total water resource pool available for consumptive use and defining rules for the allocation of that defined resource pool (hydrology considerations). In the following, we discuss Nepal's current water management development and policies as well as potential future pathways.

7.2   NATIONAL WATER RESOURCES POLICY

In order to formalize the arrangements, a draft National Water Resources Policy has been formulated (e.g., ICE WaRM, 2019; MOEWRI, 2002). This policy will guide the development of water management strategies in Nepal for the next decade, including defining the roles and responsibilities of national, provincial and local governments. The draft policy provides initial drafting instructions for complementary national water legislation which recognizes the economic implications of water and seeks equitable access for all people in the country as the country strives to achieve the global Sustainable Development Goals, especially in relation to water and community. In this context it also promotes the concept of integrated water resources management (IWRM) and the implementation of basin planning across the country.

7.2.1   Policy Goals

The following are the goals of this National Water Resources Policy (MOEWRI, 2002):

- Protection of water resources and their productive use through the medium of community involvement and participation.
- Mitigation of the adverse effects of floods, landslides and drought.
- Sufficient availability of clean water for domestic and religious/cultural uses including social and environmental uses.
- Increase in agriculture production through sufficient and reliable availability of water, and subsequent food security.
- Supply of clean electrical energy to meet domestic demand through the development of hydropower feasible from environmental, technical and economic viewpoints; increase in national income through substitution, to the extent possible, of other unclean energy uses and petroleum imports, and increase in national revenue from the export of hydropower where there is comparative advantage in doing so.

- Availability of water for other uses including industry, tourism and recreation.

7.2.2 Principles of the Policy

The Draft National Water Resources policy identifies the following principles:

- It shall be recognized that, depending upon the use and situation, water is a commodity having social and economic value, and in the course of formulating and developing projects, the cost of water resources development and management as well as the allocation of benefits from such use shall be based on the principle of 'beneficiary and polluter must pay' in accordance with the financial capacity of water users.
- The development and management of water resources shall be carried out by also considering the interrelationship between surface water and groundwater in line with the principle of integrated water resources management, in a manner to accrue optimal economic, social and environmental benefits from the available resources such as water, land and forests in each basin.
- Arrangement shall be made for the security of water and its sharing to be based on the principle of utilization by the central, provincial and local governments in accordance with the constitutional provision of the ownership of the available water resources being vested in the Nation State of Nepal and other concomitant legal provisions.
- Sustainable use shall be made of such resources by ensuring the protection of water resources and the environment, and in the case of inter-basin water transfer management shall be carried out by comprehensively considering all the concerned basins and each basin in other situations.
- While allocating water, the existing uses and equitable access shall be recognized; and the mutual obligations of consumers shall be determined for the efficient and productive use of water.
- In order to raise the necessary financial resources for future water resources development including the operation and maintenance of built infrastructure, and to maintain economy and efficiency in the use of water, appropriate tariffs, service fees, royalties, license fees, and so on, shall be levied.
- In the course of implementing a project for optimum benefit from water resources, if it impacts adversely on the existing use, the project shall be implemented only after compensating the affected consumers.
- Priority shall be given to environmental protection and maintenance of the quality of drinking water as per the constitutional provision of Nepal, and the necessary flow in terms of quantity and quality shall be maintained in order to maintain a clean and live river system; in addition, for the control of pollution, the 'polluter must pay' principle shall be adopted.

- The water distribution service shall be decentralized by involving the public, private and community organizations and autonomous entities such as users' groups; and water resources planning, design, development and management of water resources projects shall be carried out in a manner able to withstand the adverse impacts of climate change.
- Clear responsibilities shall be delineated to the concerned entities for the risk management of flood and drought, and first priority shall be given to drinking water in a situation of drought.
- A coordinated water accounting system shall be established among the various entities and consumers.
- A scientific analytical process based on the technical and socio-economic facts in order to make appropriate decisions in respect of water allocation shall be adopted.
- A gender and socially inclusive and participatory process including women and minority communities in water resources management decisions at all levels shall be adopted, water development and management related leadership shall be developed based on accountability at all government levels of the centre, province and local levels, with a clear interrelationship and responsibility among all levels.
- It shall be recognized that financial investment is of the utmost necessity for the development of infrastructure for obtaining appreciable benefits from water resources.

7.3   DISCUSSION

7.3.1   Evidence and Current State of Water Markets in Nepal

Access to water under the Constitution is considered a right, but is not currently a tradeable licence. However, several examples exist for farmer managed/informal water trading schemes, for example, in the Chherlung Thulo Kulo irrigation area (e.g., Rosegrant et al., 1995) or the Andhi Khola irrigation project (AKIP) (van Etten et al., 2002). In Chherlung Thulo Kulo, water shares were issued according to farmers' proportional contribution to the irrigation system's investment costs. Increasing opportunity costs for water and incentives for water conservation led to increased irrigated area and water use efficiency. Similarly, in the AKIP, water shares can be obtained through supplying labour or payment to build the irrigation scheme. The aim was to allocate water to all people, not on the basis of land size or land ownership status. This system, however, did not allocate water equally and poorer farmers were disadvantaged.

There is also evidence of groundwater trading. Shah et al. (2006) surveyed 20 districts of Nepal Terai and found that 91 per cent of sample villages reported water markets, and of those 67 per cent of well owners were selling water. In districts of Nepal-Sarlahi, Bhandhari and Pandey (2006) describe the groundwater market as a 'residual' market, as there are few water sellers and water is sold only when own irrigation needs are met. Water sellers typically own shallow tubewells. Usually, water buyers' farms are located adjacent to water sellers, using earthen channels for water transfers. Although this water market benefited poorer farmers, water buyers generally faced a low reliable water supply and monopolistic water pricing.

However, the general consensus is that as Nepal is one of the few countries in the world with an overall surplus of water, its focus is on the development of its water resources, particularly for hydropower or irrigation. Figure 7.2 shows the rainfall distribution across the country and it can be seen that there is a wide variation (point data were 1980–2010, taken from the Department of Hydrology and Meteorology, Kathmandu, Nepal). There are areas where water availability is limited due to the lack of rainfall and competition for access, and inter-basin transfer initiatives are currently being implemented (e.g., Bhattarai et al., 2002). In this context it is considered that rules governing these transfers will need to be developed to avoid community tensions over water allocations between basins. A nationwide programme has commenced at the Water and Energy Commission Secretariat (WECS) to undertake basin planning for all major river basins across the country. The products of this programme should identify future development options and limitations, which will further highlight the need for inter-basin transfer rules which can be considered an initial example of water trading, or at least benefit-sharing.

Given that most areas in the country are considered to have a water surplus, there is currently little pressure to adopt a water licencing and trading arrangement more universally. However, it is anticipated that the national basin planning programme (Figure 7.2 shows the major basins which are the focus of the programme) will identify a wide range of proposals suitable for development, highlighting the growing competition for available water.

7.3.2   Inter-Basin Transfer Schemes

There are certain identified inter-basin transfer schemes, which are considered to have potential to meet the irrigation demands of the water-deficit basins, namely the Babai, Bagmati, Kamala, and so on (e.g., ADB, 2015; Bhattarai, 2009). The Kathmandu Valley also has an acute water deficit, but that is mostly for the purpose of drinking water supply to an ever-increasing urban population.

The Melamchi Project, an inter-basin transfer scheme, is currently under construction to supply 170 million litres per day (MLD) of water to supplement the current water supply to Kathmandu city and ease the chronic water shortage situation within the Kathmandu Valley. Further augmentation at two stages of 170 MLD by diverting two more streams to upstream of the intake of the recently excavated 27 km long tunnel is also being considered in order to meet future demands of one of the fastest-growing urban areas in South Asia (e.g., Bhattarai et al., 2002), that was originally due for completion in 2013. In addition, a new water distribution system has been constructed for most of Kathmandu to provide water on demand to the majority of households. However, until its full implementation many areas of the city are dependent on groundwater or provision by water tankers.

Similarly, the Bheri–Babai Diversion Multipurpose Project aims to transfer 40 m³/s of water from the Bheri River to the Babai River to provide year-round irrigation water for a command area of about 51 000 ha in the Bardiya and Banke districts. The project also aims to generate power of nearly 50 MW. Tunnel-boring technology has been applied and a 12.3 km tunnel has recently been completed, under budget and well before schedule. Projects such as Bheri–Babai are important for Nepal to meet domestic and irrigation demands (GEODATA, 2019; Upadhyay and Gaudel, 2018).

7.3.3 Basin Planning and Current Water Resources Management Issues

The Kamala Basin in the southeast of Nepal has been chosen by the gov-ernment of Nepal as a pilot basin to develop capacity in basin planning. The Kamala Basin Initiative is jointly supported by the Governments of Nepal and Australia (through the auspices of the Commonwealth Scientific and Industrial Research Organisation, CSIRO) to further scientific cooperation around water resource management. The aim is to develop a strategic basin plan with relevance for other water resource-limited basins. The initiative quantifies the water availability, identifies future needs for irrigated agriculture, explores various development pathways (including participatory river basin planning) and increases ecological knowledge for sustainable management (CSIRO, 2019). The Kamala basin is important in terms of agricultural production, but has complex issues related to high levels of poverty, flooding, erosion and sediment transport, and water resources availability (peaks and troughs, as out-lined earlier). A major dam is proposed on the Sun Koshi River for diverting and supplementing water to the Kamala Basin and Irrigation District, which is currently considered to be water-deficit.

On a national basis, the WECS's countrywide Basin Planning Project is meant to be completed within the next two years and will also deliver a Master Plan for Hydropower. A draft master plan for irrigation has also been devel-oped at the Department of Water Resources and Irrigation (DWRI) and should be finalized shortly (e.g., Pradhan and Belbase, 2018). Although hydropower is a non-consumptive use of water, and the majority of the potential sites do not have competition in use with irrigation and other users, adequate attention has to be paid to ensure that conflicts do not arise, particularly in the water-deficit basins. It is desirable that wherever a choice has to be made, domestic use of water and agricultural use, including irrigation, should be ascribed first and second priorities over hydropower, as provisioned in the draft Water Resources Act (NLC, 2018b).

The possibility of conflict because of competition in water use will be further compounded if the climate change predictions identified earlier by the International Centre for Integrated Mountain Development (ICIMOD) in the Hindu Kush Himalayan Monitoring and Assessment Programme (HIMAP) report are realized (Wester et al., 2019). Significant glacier loss and greater variability of monsoonal rains throughout the region will affect future opportunities for development unless carefully planned. Decrease in snow reserves, and increases in extreme hydro-meteorological events in the form of prolonged droughts and cloudbursts, call for proper climate change resilience. Such resilience could only be achieved by bringing about changes in the use pattern, including through research into agriculture and irrigation efficiency and profitability, rather than simply by regulating the annually available water in a stream by means of storage facilities in the form of reservoirs, dams and groundwater recharge. Many potential reservoir sites are being lost due to heavy investment in other infrastructure and urbanization. Likewise, ground-water recharge is also suffering in terms of both quantity and quality. If Nepal is to achieve resilience against climate change, such potential reservoir sites and groundwater recharge areas need to be protected.

Licences are required for hydroelectricity development projects, but in many cases there are difficulties meeting community expectations for compen-sation or benefit-sharing arrangements. In several cases this community resist-ance has led to significant delays in project implementation (e.g., Ogino et al., 2019). The proposed Master Plan will identify potential projects and attempt to prioritize their development. At the time of writing in 2020, the lack of contemporary water legislation at the national and provincial level is creating some confusion and lack of accountability for water resources management (e.g., Ojha et al., 2019). Recent meetings have been held between politicians and bureaucrats from the national and provincial levels of government to consider ways to collaborate on the development of complementary legislation and management principles.

At the local level, the establishment of local governments with responsibil-ity for local water management issues has generated considerable interest in the water requirements of smaller communities, and is anticipated to increase the focus on water supply and sanitation as well as the viability of small irriga-tion areas. This focus, combined with the broader establishment of water user committees in irrigation districts, will create pressure for reform. The draft policy referred to above indicates that larger irrigation districts should be allocated water and licenced. Some form of water use agreement will be required between the district committee and individual users. It is unlikely that formal water trading arrangements will be established, due to the nature of most irrigation enterprises and the limited capacity of subsistence farmers to pay. One of the key future initiatives will be the development of national and provincial water resources legislation which ensures consistent and complementary management principles to enable secure investment in the development of Nepal's water resources for sustainable economic development and environmental protection.

Upadhyay and Gaudel (2018) provide an overview of weaknesses in Nepal's water resources management and policy planning that need consideration in the future; for example, fragmented institutions (this has improved with the formation of the Ministry of Energy, Water Resources and Irrigation, MOEWRI, in 2018), weakened institutions (for example, the WECS), multiple institutions with overlapping scope, isolated planning/decisions regarding the use of water resources, and short-term project planning. In addition, a lack of multilateral cooperation in the Ganges Basin is observed. Other studies highlight the need to update Nepal's water supply and sanitation infrastructure (Babel and Wahid, 2011) and to alleviate the growing peri-urban water competition and conflict (Shrestha et al., 2018). With the new Constitution being promulgated and the current water policy planning stage and nationwide basin planning in Nepal, it can be anticipated that lessons will be learned from the past and that current and future policies will provide the basis for sustainable water resources use and management.

7.4   CONCLUSION

Nepal is seriously considering its water regime, but is not politically or developmentally at a point to consider water trading. The current programme of basin planning to be completed in 2021 will identify available water resources and competing uses, especially for hydropower and irrigation, with Master Plans for both being completed by 2020. Thus, the application of the WMRA framework indicates that many prerequisites for effective and efficient water markets do not exist in Nepal (Suhardiman et al., 2015). With the national Basin Planning Project having just commenced, Nepal would be situated at the beginning of Step 1 of the WMRA framework (hydrology considerations and system type), quantifying the water availability of the river basins and identifying future needs for irrigated agriculture and other water users, including the environment. A future step, if formal water markets are to be considered in Nepal, is to develop appropriate institutional arrangements; for example, defining water entitlement and allocation arrangements, including monitoring and enforcement provisions.

While Nepal considers itself to be a water-surplus country, the implications of climate change may increase pressure on the availability and reliability of its water resources. Institutionally, Nepal recognizes the need to move towards a more formal water regime; but practically, given the subsistence nature of most water users, the move to formal water licences and trading, as occurs in Australia, is not considered practical at this stage. A system of water licencing/ permits for major water users is being considered, but further policy and operational development is required before it can be implemented.

REFERENCES

Asian Development Bank (ADB) (2013). *Country Environment Note – Nepal – Country Partnership Strategy: Nepal, 2013–2017*. Asian Development Bank.

Asian Development Bank (ADB) (2015). *Nepal: Innovations for More Food with Less Water. Technical Assistance Consultant's Report*. Asian Development Bank.

Babel, M.S., and Wahid, S.M. (2011). Hydrology, management and rising water vulnerability in the Ganges–Brahmaputra–Meghna River basin. *Water International*, *36*(3), 340–356.

Bhandari, H., and Pandey, S. (2006). Economics of groundwater irrigation in Nepal: some farm-level evidences. *Journal of Agricultural and Applied Economics*, *38*(1), 185–199.

Bhattarai, D. (2009). Multi-purpose projects. In D.N. Dhungel and S.B. Pun (eds), *The Nepal–India Water Relationship: Challenges* (pp. 69–98). Springer.

Bhattarai, M., Pant, D., and Molden, D. (2002). Socio-economics and hydrological impacts of intersectoral and interbasin water transfer decisions: Melamchi water transfer project in Nepal. Selected paper presented at Asian Irrigation in Transition – Responding to the Challenges Ahead, 22–23 April, Bangkok.

Commonwealth Scientific and Industrial Research Organisation (CSIRO) (2019). *Kamala Basin – Supporting River Basin Planning through the Kamala Basin Initiative*. Commonwealth Scientific and Industrial Research Organisation. https:// research.csiro.au/sdip/projects/nepal/kamala/.

Food and Agriculture Organization (FAO) (2012). *Irrigation in Southern and Eastern Asia in Figures – AQUASTAT Survey – 2011*. Food and Agriculture Organization of the United Nations.

GEODATA (2019). *Bheri Babai Diversion Multipurpose Project (BBDMP)*. GEODATA. https://www.geodata.it/en/sectors/portfolio-hydro/item/bheri-babai -diversion-multipurpose-project-bbdmp.html.

International Centre of Excellence in Water Resources Management (ICE WaRM) (2019). ICE WaRM and CSIRO support National Water Resources Policy Development in Nepal. https://www.icewarm.com.au/news/article/ice-warm-csiro -support-national-water-resources-policy-development-nepal/.

Ministry of Energy, Water Resources and Irrigation (MOEWRI) (2002). Executive summary. *Water Resources Strategy Nepal*. Ministry of Energy, Water Resources and Irrigation.

Nepal Law Commission (NLC) (2018a). *Constitution of Nepal. Part-4 Directive Principles, Policies and Obligations of the State*. Nepal Law Commission. http://www.lawcommission.gov.np/en/archives/979.

Nepal Law Commission (NLC) (2018b). *Water Resources Act, 2049 (1992)*. Nepal Law Commission. http://www.lawcommission.gov.np/en/archives/category/documents/prevailing-law/statutes-acts/water-resoures-act-2049-1992.

Ogino, K., Nakayama, M., and Sasaki, D. (2019). Domestic socioeconomic barriers to hydropower trading: evidence from Bhutan and Nepal. *Sustainability, 11*(7), 2062.

Ojha, H.R., Ghate, R., Dorji, L., Shrestha, A., Paudel, D., et al. (2019). Governance: key for environmental sustainability in the Hindu Kush Himalaya. In P. Wester, A. Mishra, A. Mukherji and A.B. Shrestha (eds), *The Hindu Kush Himalaya Assessment – Mountains, Climate Change, Sustainability and People* (pp. 545–578). Springer.

Organisation for Economic Co-operation and Development (OECD) (2003). *Development and Climate Change in Nepal: Focus on Water Resources and Hydropower*. Organisation for Economic Co-operation and Development.

Pakhtigian, E.L., Jeuland, M., Bharati, L., and Pandey, V.P. (2019). The role of hydropower in visions of water resources development for rivers of Western Nepal. *International Journal of Water Resources Development*. DOI: 10.1080/07900627.2019.1600474.

Pradhan, P., and Belbase, M. (2018). Institutional reforms in irrigation sector for sustainable agriculture water management including water users associations in Nepal. *Hydro Nepal: Journal of Water, Energy and Environment, 23*, 58–70.

Rosegrant, M.W., Schleyer, R.G., and Yadav, S.N. (1995). Water policy for efficient agricultural diversification: market-based approaches. *Food Policy, 20*(3), 203–223.

Shah, T., Singh, O.P., and Mukherji, A. (2006). Some aspects of South Asia's groundwater irrigation economy: analyses from a survey in India, Pakistan, Nepal Terai and Bangladesh. *Hydrogeology Journal, 14*(3), 286–309.

Shrestha, A., Roth, D., and Joshi, D. (2018). Flows of change: dynamic water rights and water access in peri-urban Kathmandu. *Ecology and Society, 23*(2), 42.

Shrestha, M.N. (2009). Water rights: a key to sustainable development in Nepal. *Journal of Hydrology and Meteorology, 6*(1), 37–43.

Suhardiman, D., Clement, F., and Bharati, L. (2015). Integrated water resources management in Nepal: key stakeholders' perceptions and lessons learned. *International Journal of Water Resources Development, 31*(2), 284–300.

Upadhyay, S., and Gaudel, P. (2018). Water resources development in Nepal: myths and realities. *Hydro Nepal: Journal of Water, Energy and Environment, 23*, 22–29.

van Etten, J., van Koppen, B., and Pun, S. (2002). Do equal land and water rights benefit the poor? Targeted irrigation development: the case of the Andhi Khola Irrigation Scheme in Nepal. IWMI Working Papers H030201, International Water Management Institute (IWMI).

Water and Energy Commission Secretariat (WECS) (2011). *Water Resources of Nepal in the Context of Climate Change*. Water and Energy Commission Secretariat.

WECS and CSIRO (2019). *State of the Kamala River Basin, Nepal*. 68pp.

Wester, P., Mishra, A., Mukherji, A., and Shrestha A.B. (2019). *The Hindu Kush Himalaya Assessment – Mountains, Climate Change, Sustainability and People*. Springer.

Wheeler, S.A. (2014). Insights, lessons and benefits from improved regional water security and integration in Australia. *Water Resources and Economics, 8*, 57–78.

Wheeler, S.A., Loch, A., Crase, L., Young, M., and Grafton, R. (2017). Developing a water market readiness assessment framework. *Journal of Hydrology*, *552*, 807–820.

World Bank (2019). *Nepal*. World Bank. https://data.worldbank.org/country/nepal.

8. Groundwater markets in the Indus Basin Irrigation System, Pakistan

Irfan Ahmad Baig, Muhammad Ashfaq and Rida Afzal

8.1 INTRODUCTION AND BACKGROUND

Water is no doubt one of the most critical environmental resources for sustaining life on earth. Since the last century, the world has experienced constantly expanding demand for water, owing to a rapidly increasing population and changing lifestyles (Qu et al., 2013). Water scarcity in many countries is increasing by about 1 per cent annually and is expected to increase at the same rate over the next couple of decades, with regularly occurring droughts posing a serious socio-economic threat to societies, particularly in developing countries (WWAP, 2019). The increasing scarcity of water is emerging as an alarming threat for socio-economic prosperity, geopolitical stability across the globe and the violation of fundamental human rights (Harris et al., 2015; UN Water, 2015).

Water security threats have become an even graver issue within the changing climate, especially for drier regions across the globe, which are becoming increasingly dry due to erratic rainfall and precipitation. This issue has also created multiple challenges for water apportionment and cross-border water disputes (Tsur, 2009). In a world where water shortages are expected to increase about 50 per cent by 2050 under a business-as-usual scenario, paving the way for sustainable development will be crucial (Ako et al., 2010). Inefficient water use practices have already created a huge gap in demand and supply for water, which is further exacerbated due to the fragmented, divergent and vast landscape in developing countries, such as Pakistan (Everard et al., 2018), thus increasing the need for water markets to cope with water scarcity and improve equity (Razzaq et al., 2019). However, factors such as poor infrastructure, socio-political systems driven by contradictions, deteriorated water quality and skewed land distribution often do not support the operation of water markets in such countries (Wheeler et al., 2017).

The development of water markets is often regarded as an advantageous economic instrument to mitigate increasing water scarcity and to promote equity among water users. Water markets are seen by policy-makers and water managers as a means to achieve economic and environmental goals through efficient water allocation and use, when technical approaches to expand water resources are not economically or politically viable. In contrast to administrative mechanisms, water markets have shown to be effective with regard to efficient resource allocation (Garrido and Calatrava, 2010), as water prices determined through water markets make the opportunity cost of water clear to the users (Qureshi et al., 2009). Water markets have emerged as a coping tool to control water scarcity and drought in countries like Australia (Qureshi et al., 2009; Bjornlund and Rossini, 2010). However, in contrast to the United States (US) and Australia, water markets in Asian countries have small irrigators involved in groundwater exchanges, which are informal in nature and trade in limited amounts of water (Zhang et al., 2008).

In a developing country such as Pakistan, generous subsidies, inefficient pricing and the marketing mechanism of surface water (Mekonnen et al., 2016) have added to the unreliability of surface water, rendering buyers little control. On the other hand, various studies have reported greater reliability of groundwater and greater control by users. In Pakistan, groundwater markets are informal in nature; they are not officially recognized, but are gradually expanding across Pakistan and are considered as a mechanism for informal water trade. However, they do not involve the exchange of permanent water entitlements (Khair et al., 2011). We explore this further with a case study of the Indus Basin in Pakistan.

8.2 CASE STUDY

8.2.1 Indus Basin, Pakistan

The River Indus is one of the major rivers within Asia, and stretches across the majority of Pakistan (Hassan, 2016). The Indus River Basin has a total area of 1.12 million km^2, from the Himalayan Mountains in the north to the dehydrated terrestrial plains of Sindh province in the south, where it discharges into the Arabian Sea. The Indus Basin Irrigation System (IBIS) is distributed across four countries: Pakistan, India, China and Afghanistan. In Pakistan, the Indus River covers an area of 520 000 km^2, or 65 per cent of its territory, comprising Khyber Pakhtunkhwa (KP), Punjab and Sindh provinces (FAO, 2011). See Table 8.1.

Over 1.7 billion inhabitants currently located in the Indus Basin rely on the water from river flows or groundwater, which uses natural water recharge, supported mainly through rainfall and river flows. The same river system is

Table 8.1   Area distribution of Indus River Basin

Basin	Area (million km[2)]	Countries' share Indus River	Area of country in basin (km[2)]	% of area of IBIS	As a % of total area of the country
IBIS	1.12	Pakistan	520 000	47	65
		India	440 000	39	14
		China	88 000	8	1
		Afghanistan	72 000	6	11

Source: FAO (2011).

also being used for the disposal of over 80 per cent of human waste and, unfortunately, because of lack of treatment, 663 million people still lack access to safe drinking water (Young et al., 2019).

8.2.2   Water Resources and their Development in the Indus Basin

About 50 per cent of the water within the Indus Basin originates from glaciers, while the rest comes from seasonal rainfall, snow, summer streams and runoff (Adams, 2019). The main annual discharge in the Indus basin is $6.2 \times 10^{10}\,\text{m}^3$ (Amin et al., 2018). The climate is complex across the basin because it ranges from subtropical to arid and temperate. This area endures varying seasonal rainfall patterns, timing of snow and glacier melting, high river flows before the productive season, and dry streams over the whole of summer (Mesquita et al., 2019).

Surface water resources in Pakistan are based on the flow of the Indus River and its tributaries (Jhelum, Chenab, Ravi, Sutlej and Beas), and the inflow to these rivers derives from snow, glacial melt and rainfall in catchment areas (Qureshi, 2011). During the first decade of the 21st century, a sharp decline in surface water resulted from climatic variations (high runoff, unreliable precipitation and extended drought), which instigated farmers to exploit more groundwater. A 26 per cent reduction in surface water flows made groundwater the last resort for irrigation and drinking water, and resulted in a 59 per cent increase in the growth of private tube-wells (Qureshi et al., 2009). At present, more than 1 million tube-wells are operating across the Indus Basin of Pakistan.

8.3   PAKISTAN: DEPENDENCE ON THE INDUS BASIN IRRIGATION SYSTEM (IBIS)

Pakistan's agriculture relies heavily on the irrigation network of the IBIS, critical for its significant role in agricultural production – yet not limited to agriculture – given the network supplies water across most sectors of the economy. The Basin has two multi-purpose reservoirs, 19 barrages, 12 inter-river link canals, and 45 irrigation canals that cover 18 million hectares and provide 120 000 water sources to farms. The IBIS supports 90 per cent of food production in the country (Yang et al., 2013).

Pakistan's predominantly agrarian society is economically and socially dependent on the Indus River System (IRS) with its links to agriculture and industry. Irrigated agriculture contributes more than 90 per cent of food production. Each year, changing weather, water and agro-economic conditions across the basin create unique circumstances for management (Yu et al., 2013). In Pakistan, an irrigated area of around 21.45 Mha devoted to intensive production of food crops was impacted upon by dwindling prices and unfavourable weather conditions (FAO, 2017).

Considering Pakistan's arid and semi-arid climate, agriculture predominantly relies on irrigation from both canal and groundwater, and any gap in the supply and demand of irrigation water affects agriculture, industry or domestic consumers simultaneously, due to the contiguous nature of the network (Amin et al., 2018). Owing to increasingly variable and erratic rainfall, surface water supplies are highly unreliable in the absence of sufficient storage reservoirs in the Indus River System, thus increasing the risk of water shortages for agriculture and human consumption (Alamgir et al., 2016). Due to rapid development and urbanization, there has been increased competition for water resources (Salik et al., 2016). A further problem (from a purely private irrigation perspective) is conveyancing losses; Table 8.2 demonstrates that almost 60 per cent of water is 'lost' across the Indus Basin system. However, it must also be noted that much of this loss is actually a benefit for environmental water use and a recharge back to surface and groundwater sources (Grafton et al., 2018).

Water rights are not linked to the productivity of water, but to land ownership in Pakistan. Many studies conducted in different agro-ecological zones have estimated that Pakistan is ranked last for water productivity of crops. A recent study (Nawaz, 2018) compared the productivity among different countries, reflecting that Pakistan has the lowest crop-water productivity in the region (130 gram/m^3 in comparison with 390 and 800 gram/m^3 in India and China, respectively). The average water productivity of rice is 0.45 kg/m^3 compared to the world average of 0.71 kg/ m^3.

8.4   IRRIGATION WATER SCARCITY

Surface water is the main source of irrigation in Pakistan, with a storage capacity of 140 million acre-feet (MAF). It provides canal withdrawal of 104 MAF and comprises three dams (Tarbela, Mangla and Chashma), nine barrages, 55 canals and 1.621 million km of extended water channels (Hussain

Table 8.2   Irrigation water losses across the IBIS

Category	Supply (MAF)	Losses (MAF)	%
Average flows	144		
Canal diversion	105		
Canal losses		26	25
Canal supply at watercourse head	79		
Watercourse losses		24	30
Available at farm gate	55		
Groundwater contribution	50		
Total available	105		
Field channel losses		11	10
Available at field level	95		
Field application losses		25	26
Available at crop level	70		
Rainfall	13		
Total available for crops	83		

Sources: Adapted from Bhatti et al. (2017) and Stewart et al. (2018).

and Mumtaz, 2014). Recent estimates show that surface water availability has declined around 15 per cent over past decades, but has also become inaccessible in many parts of the Basin, turning it into one of the most depleted river basins in the world (Watto and Mugera, 2016). Abnormally high seasonal fluctuations, heavily melting glaciers and ice, large quantities of water lost as runoff, are all important factors for surface unreliability and seasonal scarcity (Salik et al., 2016). Water shortages are primarily due to a range of factors, including:

- Climate change. According to the Fifth Assessment Report (AR5) of the Intergovernmental Panel on Climate Change (IPCC), climate change will have serious consequences for the availability and properties of natural water reserves, especially in Pakistan. Due to global warming, hydrological systems will alter water availability, specifically due to melting of Upper Indus glaciers that contribute more than 50 per cent to the river supply (Muhammad et al., 2019; Yaqoob, 2016).
- Seasonal variation. The temperature of the Indus Basin changes seasonally. Low winter runoff and high summer runoff are common. Hydrological studies carried out in the Sutlej River found that the flow variations are due to seasonal changes. In winter, low runoff creates water deficiency

for agricultural crops, and during the high run-off season a lot of water is unused, without any significant positive impact on agricultural productivity because of insufficient water storage (Hussain and Mumtaz, 2014).

- Water quality. The Indus River Basin is facing severe water quality degradation (Young et al., 2019) due to excessive use of fertilizer and improper disposal of waste (Qureshi, 2011; Wang et al., 2019). During recent decades, indiscriminate water extraction has lowered the water tables and encouraged the intrusion of brackish water (Bhatti et al., 2017). Livelihoods substantially dependent on fisheries and agriculture remain extremely vulnerable to increasing salinity (Salik et al., 2016). Conservative estimates show that less than 20 per cent of the area has safe groundwater (PCRWR, 2008, 2016). In addition, the concentration of arsenic has exceeded World Health Organization (WHO) limits in most parts of Pakistan, causing many problems, such as skin diseases in humans, either by direct contact with water or by consuming agricultural products (Hussain et al., 2016). This reduces water availability for beneficial use and increases its cost of consumption (FAO, 2017).
- Limited storage capacity and poor infrastructure. Pakistan has limited water storage capacity: 15 per cent of river flow per year. Per capita water storage potential in Pakistan is 50 m^3, compared to 5000 m^3 in the US, and China's 2200 m^3. Pakistan can store water for barely 30 days, compared to 100–120 days in India and 500 days in South Africa. Pakistan's water sector strategy concluded that there is a need to increase storage potential by 22 billion cubic metres (BCM) by 2025 to deal with the projected requirement of 165 BCM. It is due to poor management that, even after the establishment of the Tarbela dam, no decision was taken to construct further infrastructure to increase storage capacity (Qureshi, 2011).
- Widening of supply and demand gap. Water demand for agricultural, commercial and domestic uses is projected to increase 53 per cent by 2025 compared to 2010, while water availability is expected to fall up to 800m^3 (Qureshi, 2011; Raza et al., 2009; Amir and Habib, 2015). Land per person for agricultural production has also reduced, and agriculture is threatened by waterlogging and increasing salinity (Qureshi, 2011).

8.5 PAKISTAN: THE CONTRIBUTION OF GROUNDWATER

In Pakistan, groundwater pumping on a larger scale started in the 1960s, thus decreasing the water table in the Indus Basin, which had built up due to seepage of freshwater from canals (illustrating that many of the losses from diversions, as described earlier, are not true losses). In the 1970s, the government supported farmers by subsidizing electricity (Qureshi, 2016). In the

1980s, the uneven growth in pumping increased and over-extraction occurred. The government introduced a licensing system to restrict the installation of private tube-wells in a few areas where groundwater levels had been falling sharply (Mekonnen et al., 2016). In Punjab and Sindh provinces, indirect management/restrictions were applied by removing the energy subsidies. However, this resulted in an enormous increase in the count of fuel-operated water pumps and engines, which had low installation costs, were suitable for small farms and gave irrigators more control over farming operations (Mekonnen et al., 2016). Many studies concluded that farmers pump groundwater because of its reliability and control of use; in contrast to canal water, which has an unreliable supply and does not always match demand (Kazmi et al., 2012; Ahmad, 2016). However, in Pakistan groundwater has traditionally been used to fill the gap between the demand and supply of surface water for irrigation. The preference for surface water is mainly due to its existing price structure, which is very low (almost free), and its perceived better water quality. Dependence on surface water decreased as irrigation infrastructure and the water conveyance system lost their effectiveness – mainly due to poor operation and maintenance, an over-committed system, distribution and application inefficiencies, and low levels of cost recovery – while irrigation water demand has been rising due to increased food demand and industrial pressure (Baig, 2009; Brooks and Harris, 2014). Currently, more than 50 per cent of the total crop water requirement is supplemented by groundwater resources, via the use of over 1 million tube-wells across Pakistan.

8.6   WATER MARKET DEVELOPMENT IN THE INDUS BASIN OF PAKISTAN

In Pakistan, informal water markets are generally common. For example, 60 per cent of farmers purchased water in return for a crop share, while almost 40 per cent trade water as a cash transaction (Doll et al., 2012). Trade in water markets depends on location of the farm, water table depth and water price per hour. Water price varies with the depth of the water table: as it moves from a deep to high water table, the price lowers (Khair et al., 2012). However the significant effect of these informal markets is that they offer opportunities to the well owners for significant financial benefits, and for non-owners to increase productivity (Manjunatha et al., 2011).

One drawback of this type of informal market is its free access to water in Pakistan, leading to consumptive losses. In contrast to surface water, groundwater certification is not determined. Access to groundwater for irrigation is free and bound to land ownership, and access to water is not defined nor restricted. Poor policies and governance cause misallocation of water. A tube-well owner can extract, and even sell, water without any official certification (Qureshi et al., 2010a). The factors that affect water markets are not only physical factors – such as cropping patterns, water quality, alternative sources of irrigation – but also knowledge, awareness, education and kinship (Khair et al., 2012).

8.6.1 Groundwater Markets in Pakistan

As discussed above, surface water in Pakistan – due to its extremely low price – has been the first choice of farmers. However, over time, as this resource has depleted, farmers have come to rely on groundwater, which may be owned or purchased. The ownership of tube-wells is dependent on the size of land-holdings and the financial position of farmers: the majority of these wells are owned by farmers with medium-sized and large properties, while non-owners are mostly small landholders. Since there are large numbers of non-owners who need irrigation water, they have to purchase water from others.

In fact, groundwater marketing is increasing the availability of irrigation water for farmers as well as for production of crops. Despite the advantage of these markets, it is also clear that the degree of control of water affordability to purchasers is not as much as that of tube-well owners. Tube-well owners treat water sales as a residual category, to be met after serving the needs of their own fields. The extra water is sold while considering many other social preferences by both sides of the marketing chain (cast system, social relations, and so on).

8.6.2 Groundwater Marketing: Water Buyers and Sellers

Under tube-well-based farming in much of south Asia, water that may be accessed by a farmer to meet the needs of their own farm is often sold to nearby farmers. Farmers of all categories (small, medium, large) sell water. The groups most involved in selling water were local dominant groups who are strong and rich in terms of landholding and social status. Access to alternative irrigation sources also affects groundwater selling. Groundwater selling increases where access to surface water is low. Moreover, the incidence of water sales was higher in wet rather than dry areas, because wells in dry areas yield less than those in wet areas (Khair, 2013).

Farmers who are willing to trade water are mostly large commercial farmers who produce at a large scale and generally have better access to surface water; thus they can afford to trade water, depending upon their needs and their ability to pay better prices. Similarly, farmers with large tube-wells desire to sell water to add to their seasonal revenue, when they are getting sufficiently large quantities of surface water. Moreover, farmers who have greater flexibility in the irrigation industry can sell water to compensate if they face any issues during their use of the water (Razzaq et al., 2019).

Another factor that has traditionally affected non-formal groundwater markets is the ability of farmers to sell water. Recent research has reflected that, in local contexts, fragmentation of land, crop rotation, social pressure and the acceptability of the water deal (due to existing pressures) significantly affect the ability of farmers to sell water in different contexts and seasons.

The decision to buy water is mainly driven by farm size (small, medium or large) and the farm's water quality. The size of land ownership had a significant negative effect on water purchasing decisions (Meinzen-Dick and Sullins, 1994; Manjunatha et al., 2011). Several climatic and non-climatic variables also influence the buying of water. Rain is the most important factor in water buying decisions (Khair, 2013). Different projections show that, in the absence of effective demand management, water supply will not be able to meet rising water requirements, especially in the context of escalating temperatures. Water buyers will have to pay higher prices, and an improvement in water use efficiency will need to occur.

8.6.3   Limitations to Groundwater Marketing

Traditionally, in Pakistan groundwater has been considered a substitute for surface water; however, its entitlements are not clear. Access to groundwater for irrigation is open and is commonly linked with land ownership. There is no restriction in terms of policy, governance or market regulations regarding groundwater consumption and allocation.

In local settings, where a well owner has exclusive rights to extract and utilize groundwater for irrigation, urban or industrial use, through either socially optimal or suboptimal ways, they can extract, or even sell water without any interference (Qureshi et al., 2010b). Such informal transactions take place through locally established and traditionally governed groundwater markets. These informal markets also provide opportunities for tube-well owners to gain high economic benefits, while also allowing non-owners to raise levels of agricultural productivity (Manjunatha et al., 2011). However, such transactions usually do not consider the shadow price of groundwater.

Recent energy crises and lowering groundwater tables in many irrigated areas have complicated the nature and functioning of informal groundwater markets. A water buyer must pay water charges in advance and let the well owner know about their water demand well in advance in order to buy water. Such situations have also led towards a time scheduling of water supplies, where well owners must develop a schedule for water supplies through mutual agreement. The closer the water buyer's farm, the more immediate will be their approach to water. In areas with thin density of tube-wells, but high dependency on groundwater, water buyers have no or limited options to select among a number of sellers, thus letting well owners exercise some monopoly power (Watto, 2015).

Similarly, unlimited access to this precious resource has also permitted tube-well owners in many areas to enter markets in such a way that they use their groundwater resources only for selling, without even having any considerable area of land for cultivation. They provide their groundwater at a discounted price, conditional on the water buyer using their own tractor or diesel engine to extract the water. This practice leads to inefficient use of groundwater in irrigation, and also raises equity concerns. Social ties between water sellers and buyers can cause discrimination (Watto, 2015).

Groundwater markets are significantly determined by the demand–supply gap of surface water, but also by water pricing. The long history of surface water charges in Pakistan is characterized by heavy subsidies; thus, cost recovery through water charges was not even sufficient to cover the operating and maintenance charges of the system. Water prices charged are 14–90 times higher than operating and maintenance costs of running a tube-well. Different initiatives by the government attempted to improve the cost recovery ratio, but lack of effective implementation of such initiatives has always remained a constraint (Watto, 2015).

The current charging system of irrigation water services does not enhance water use efficiency, because it fails to describe the association between the water price and the actual water volume provided to the farm (Qamar et al., 2018). Qureshi et al. (2003) compared the water cost from surface and groundwater sources, depicting a huge difference in water rates. The cost of water also varies with different types of water service or pumping. The minimum price of water can be set by including operating and maintenance factors within the cost, while the recovery of capital costs, partially or completely, is a policy decision. The operating and maintenance costs, currently assessed at approximately Rs300/acre (about US$4.7/ha), need to be recovered from water consumers (Nawaz, 2018).

8.7 WATER GOVERNANCE

In Pakistan, the water sector has been subject to overlapping responsibilities, with no clear division of roles and responsibilities (Cooper, 2018). Although water distribution and maintenance of the system is largely a provincial responsibility, the coexistence of a large number of federal institutions and policies overlap the provincial systems and policies, thus creating a complex patchwork of policies (Young et al., 2019). In relation to international transboundary water governance issues, Pakistan is a signatory to the Indus Water Treaty (IWT), signed in 1960. Pakistan's Constitution guarantees the right to water for every citizen through Article 9. The federal government takes care of two areas: interstate water disputes and policy settings for water and power development, mandated through the Water and Power Development Authority Act (1958). Currently, water distribution to the provinces is carried out according to the Indus Apportionment Accord 1991 and the Indus River System Authority Act 1992.

Provinces are responsible for distribution of water to the end users through a system of predefined distribution schedules known as 'Warabandi'. Traditionally, the distribution system is based on the water supply system and is regulated through Provincial Irrigation and Drainage Acts. A notable feature of this system is the significant involvement of subsidies, implied policy, implicit laws and rules and engineering-based administration. Some modification has been made; however, the quality of overall water supply services continued to decline (Baig, 2009). In 1997, the government of Pakistan tried to involve local communities and end-users in the management of the system, by establishing Irrigation and Drainage Authorities (PIDAs) at provincial levels. PIDAs provided a management platform for system users and were entrusted with managing the irrigation distribution systems within a largely decentralized governance architecture. However, different organizations created under the PIDA system could not sustain this system, because they were considered a parallel system to the publicly managed one (WWF, 2012; Young et al., 2019). In 2019, the PIDA system was rolled back in the Punjab in the name of instituting a more dynamic system for irrigation management.

The government of Pakistan has also devised independent explicit water supply policies administered by local government and intended to promote new legislative infrastructure; however, they also have limitations. A better outcome could be achieved by negotiation with local stakeholders. The respective governments and departments need to create an enabling environment and find legitimate instruments for implementing water policies (Nawab and Nyborg, 2009). An important aspect of water laws in Pakistan is the role of local governments in management of groundwater in the cities, because they are entitled to control private sources of groundwater supply in urban areas.

The legislative framework for water management in Pakistan was the Irrigation and Drainage Act of 1973. The Soil Reclamation Act was approved in 1952. This provides authority to combat water logging and salinity using tube-wells, and farmers were allowed to install private tube-wells (Watto, 2015). In 1958, the Water and Power Development Authority (WAPDA) was constituted and entrusted with the task of water allocation, becoming part of a research and development portfolio after 1970. WAPDA also had the control over Pakistan's groundwater resources under the WAPDA Act, and can set official area-specified rules, such as water licensing systems for tube-well installation. The concentration of power into this single authority started to reveal complications in the shape of growing concerns and conflicts among provinces and different players in the system.

In response to rapidly declining groundwater levels, in 1978 a groundwater rights administration ordinance was created for issuance of licenses for tube-well installation. Local district water committees were established. One of the features of these ordinances was an area-based license. These licenses were rarely implemented, because surface water became increasingly scarce over time, with a greater dependence on groundwater. Similarly, many other Acts and regulations, such as the Provincial Irrigation and Drainage Authority Act (1990–2000) and the Canal Act of 2006, remain pending (Watto, 2015).

Hence, institutions created to prevent unsustainable practices did not function effectively, and Pakistan is trying to reform institutions, policies and practices (Mustafa et al., 2013). However, the sheer number of institutions at federal, provincial and local levels (ten public sector institutions, 28 national organizations and 19 academies) to cover the water sector and to govern surface water distribution in Pakistan is said to hinder reform (Kamal, 2009). In the groundwater sector, the network of institutions is limited only to the supply of extracted groundwater to the urban/household sector; whereas irrigation groundwater supplies are still largely unregulated. This has created the potential for converting informal partnerships between multiple stakeholders for the use of water resources.

8.8   CONCLUSION

Water allocation and supply in the Indus Basin is facing multiple challenges, including increasing urbanization, population pressure, social complexities, inefficient application in agriculture, distorted prices, political conflicts and cross-border rifts. These challenges have become critical with the onset of deteriorating climatic conditions, particularly over the last two decades. Unreliable surface water supplies, where water rights are bundled with land, have put enormous pressure on groundwater extraction.

Water governance and market-based instruments could provide a management framework leading towards sustainable water use, especially if groundwater is considered. However, a review of the existing water institutions and management structure revealed the inherent inability of the system to adjust to the changing realities of the water sector. Even the 'Stage 1' requirement for the presence of efficient water markets of Wheeler et al. (2017) – that is, knowledge of the hydrological system – is not effectively in place. It is also evident that the establishment of multiple institutions in the water sector has only added to the complexities, where water, as a resource, is not being treated as an economic good. There is wider scope for establishment of public–private partnerships and improved forms of participatory irrigation management involving stakeholders at various levels for instigating political, economic and financial sustainability, and enhanced efficiency in the water sector. Now is the time to improve historic concepts of water allocation, knowledge of the system, operational methods and planning, to treat water users as consumers by efficient allocation of the resource.

REFERENCES

Adams, T.E. (2019). Water resources forecasting within the Indus River Basin: a call for comprehensive modelling. In T.E.A. Adams and S.I. Khan (eds), *Indus River Basin: Water Security and Sustainability* (pp. 267–308). Elsevier.

Ahmad, S. (2016). Water sector of Pakistan: a situational analysis. *Development Advocate Pakistan, 3*(4), UNDP, Pakistan. https://www.undp.org/content/dam/pakistan/docs/DevelopmentPolicy/DAP%20Volume3,%20Issue4%20English.pdf.

Ako, A.A., Eyong, G.E.T., and Nkeng, G.E. (2010). Water resources management and integrated water resources management (IWRM) in Cameroon. *Water Resources Management, 24*(5), 871–888.

Alamgir, A., Khan, M.A., Manino, I., Shaukat, S.S., and Shahab, S. (2016). Vulnerability to climate change of surface water resources of coastal areas of Sindh, Pakistan. *Desalination and Water Treatment, 57*(40), 18668–18678.

Amin, A., Iqbal, J., Asghar, A., and Ribbe, L. (2018). Analysis of current and future water demands in the Upper Indus Basin under IPCC climate and socio-economic scenarios using a hydro-economic WEAP Model. *Water (Switzerland), 10*(5), https://doi.org/10.3390/w10050537.

Amir, P., and Habib, Z. (2015). *Estimating the Impacts of Climate Change on Sectoral Water Demand in Pakistan.* Action on Climate Today. https://cyphynets.lums.edu.pk/images/Readings_concluding.pdf.

Baig, I.A. (2009). An analysis of irrigation charges and cost recovery under the reforms era: a case study of Punjab, Pakistan. University of Agriculture, Faisalabad, Pakistan.

Bhatti, M.T., Anwar, A.A., and Aslam, M. (2017). Groundwater monitoring and management: status and options in Pakistan. *Computers and Electronics in Agriculture, 13*, 143–153.

Bjornlund, H., and Rossini, P. (2010). Climate change, water scarcity and water markets – implications for farmers' wealth and farmer succession. 16th Annual Conference of the Pacific Rim Real Estate Society, Wellington, New Zealand, 24–27 January.

Brooks, R., and Harris, E. (2014). Price leadership and information transmission in Australian water allocation markets. *Agricultural Water Management, 145*, 83–91.

Cooper, R. (2018). *Water Management/Governance Systems in Pakistan.* K4D Helpdesk Report. Institute of Development Studies.

Doll, P., Hoffmann-Dobrev, H., Portmann, F.T., Siebert, S., Eicker, A., et al. (2012). Impact of water withdrawals from groundwater and surface water on continental water storage variations. *Journal of Geodynamics, 59–60*, 143–156.

Everard, M., Sharma, O.P., Vishwakarma, V.K., Khandal, D., Sahu, Y.K., et al. (2018). Assessing the feasibility of integrating ecosystem-based with engineered water resource governance and management for water security in semi-arid landscapes: a case study in the Banas catchment, Rajasthan, India. *Science of the Total Environment, 612*, 1249–1265.

FAO (2011). *Indus Basin River. Irrigation in Southern and Eastern Asia in Figures – AQUASTAT Survey – 2011.* Food and Agriculture Organization.

FAO (2017). *Water for Sustainable Food and Agriculture.* Food and Agriculture Organization.

Garrido, A., and Calatrava, J. (2010). *Agricultural Water Pricing: EU and Mexico.* Organisation for Economic Co-operation and Development. OECD Study on Sustainable Management of Water Resources in Agriculture, 47 pp.

Grafton, R.Q., Williams, J., Perry, C.J., Molle, F., Ringler, C., et al. (2018). The paradox of irrigation efficiency. *Science, 361*(6404), 748–750.

Harris, L.M., Rodina, L., and Morinville, C. (2015). Revisiting the human right to water from an environmental justice lens. *Politics, Groups, and Identities, 3*(4), 660–665.

Hassan, M. (2016). Water security in Pakistan: issues and challenges. United Nations Development Programme Pakistan, *Development Advocate Pakistan, 3*(4). https:// www.undp.org/content/dam/pakistan/docs/DevelopmentPolicy/DAP%20Volume3, %20Issue4%20English.pdf.

Hussain, M., and Mumtaz, S. (2014). Climate change and managing water crisis: Pakistan's perspective. *Reviews on Environmental Health, 29,* 71–77.

Hussain, Y., Dilawar, A., Ullah, S.F., Akhter, G., Martínez, H., et al. (2016). Modelling the spatial distribution of arsenic in water and its correlation with public health, central Indus Basin, Pakistan. *Journal of Geoscience and Environment Protection, 4*(18). DOI: 10.4236/gep.2016.42003.

Kamal, S. (2009). *Pakistan's Water Challenges: Entitlement, Access, Efficiency and Equity.* Woodrow Wilson International Centre for Scholars.

Kazmi, S.I., Ertsen, M.W., and Asi, M.R (2012). The impact of conjunctive use of canal and tube well water in Lagar irrigated area, Pakistan. *Physics and Chemistry of the Earth, 47,* 86–98.

Khair, M.S. (2013). The efficacy of groundwater markets on agricultural productivity and resource use sustainability: evidence from the upland Balochistan region of Pakistan. Unpublished PhD thesis, Charles Sturt University, Wagga Wagga, NSW.

Khair, S.M., Mushtaq, S., Culas, R., and Hafeez, M. (2012). Groundwater markets under the water scarcity and declining watertable conditions: the upland Balochistan Region of Pakistan. *Agricultural Systems, 107*(C), 21–32.

Khair, Syed M., Mushtaq, S., Culas, R.J., and Mohsin, H. (2011) Groundwater markets under the water scarcity conditions: the upland Balochistan region of Pakistan. 40th Australian Conference of Economists (ACE 2011), 11–13 July, Canberra, Australia. https://eprints.usq.edu.au/19453/.

Manjunatha, A.V., Speelman, S., Chandrakanth, M.G., and Van Huylenbroeck, G. (2011). Impact of groundwater markets in India on water use efficiency: a data envelopment analysis approach. *Journal of Environmental Management, 92*(11), 2924–2929.

Meinzen-Dick, R.S., and Sullins, M. (1994) Water markets in Pakistan: participation and productivity. EPTD Discussion Paper 4. http://ageconsearch.umn.edu/record/ 42824.

Mekonnen, D., Siddiqi, A., and Ringler, C. (2016). Drivers of groundwater use and technical efficiency of groundwater, canal water, and conjunctive use in Pakistan's Indus Basin Irrigation System. *International Journal of Water Resources Development, 32*(3), 459–476.

Mesquita, M.D.S., Orsolini, Y.J., Pal, I., Veldore, V., Li, L., et al. (2019). Challenges in forecasting water resources of the Indus River Basin: Lessons from the analysis and modeling of atmospheric and hydrological processes. *Indus River Basin: Water*

Security and Sustainability, 57–83. DOI: https://doi.org/10.1016/B978-0-12-812782 -7.00003-5.

Muhammad, S., Tian, L., and Khan, A. (2019). Early twenty-first century glacier mass losses in the Indus Basin constrained by density assumptions. *Journal of Hydrology, 574*(March), 467–475.

Mustafa, D., Akhter, M., and Nasrallah, N. (2013). *Understanding Pakistan's Water– Security Nexus.* United States Institute of Peace.

Nawab, B., and Nyborg, I.L.P. (2009). Institutional challenges in water supply and sanitation in Pakistan: revealing the gap between national policy and local experience. *Water Policy, 11*(5), 582–597.

Nawaz, M. (2018). Water market in Pakistan – a case for revenue generation and water security. International Symposium on 'Creating a Water Secure Pakistan', Islamabad, Pakistan.

PCRWR (2008). *National Water Monitoring Program 2002–06.* Government of Pakistan, Ministry of Science and Technology, Pakistan Council of Research in Water Resources (PCRWR).

PCRWR (2016). *Water Quality Status of Major Cities of Pakistan 2015–2016.* Government of Pakistan, Ministry of Science and Technology, Pakistan Council of Research in Water Resources (PCRWR).

Qamar, M.U., Azmat, M., Abbas, A., Usman, M., Shahid, M.A., and Khan, Z.M. (2018). Water pricing and implementation strategies for the sustainability of an irrigation system: a case study within the command area of the Rakh branch canal. *Water (Switzerland), 10*(4), 1–24.

Qu, S., Gong, M., and Dong, H. (2013). A water management strategy based on efficient prediction and resource allocation. *IERI Procedia, 4,* 224–230.

Qureshi, A.S (2011). Water management in Indus basin in Pakistan: challenges and opportunities. *Mountain Research and Development, 31,* 252–260.

Qureshi, A.S. (2016). Perspectives for bio-management of salt-affected and waterlogged soils in Pakistan. In J. Dagar and P. Minhas (eds), *Agroforestry for the Management of Waterlogged Saline Soils and Poor-Quality Waters.* Advances in Agroforestry, Vol. 13. Springer. https://doi.org/10.1007/978-81-322-2659-8_6.

Qureshi, A.S., Gill, M.A., and Sarwar, A. (2010a). Sustainable groundwater management in Pakistan: challenges and pportunities. *Irrigation and Drainage, 59*(2), 107–116.

Qureshi, A.S., McCornick, P.G., Sarwar, A., and Sharma, B.R. (2010b). Challenges and prospects of sustainable groundwater management in the Indus Basin, Pakistan. *Water Resources Management, 24*(8), 1551–1569.

Qureshi, A.S., Shah, T., and Akhtar, M. (2003). The groundwater economy of Pakistan. Working Paper No. 64, IWMI, Colombo, Sri Lanka.

Qureshi, M.E., Shi, T., Qureshi, E.S., and Proctor, W. (2009). Removing barriers to facilitate efficient water markets in the Murray–Darling Basin of Australia. *Agricultural Water Management, 96*(11), 1641–1651.

Raza, M.A., Ashfaq, M., Zafar, M.I., and Baig, I.A. (2009). Impact assessment of institutional reforms in irrigation sector on rice productivity: a case study of Punjab, Pakistan. *Pakistan Journal of Life and Social Sciences, 7*(1), 16–20.

Razzaq, A., Qing, P., Naseer, M., Abid, M., Anwar, M., and Javed, I. (2019). Can the informal groundwater markets improve water use efficiency and equity? Evidence from a semi-arid region of Pakistan. *Science of the Total Environment, 666,* 849–857.

Salik, K.M., Hashmi, M., Ishfaq, S., and Zahdi, W. (2016). Environmental flow requirements and impacts of climate change-induced river flow changes on ecology of the Indus Delta, Pakistan. *Regional Studies in Marine Science, 7*(June), 185–195.

Stewart, J.P., Podger, G.M., Ahmad, M.D., Shah, M.A., Bodla, H., et al. (2018). *Indus River System Model (IRSM) – A Planning Tool to Explore Water Management Options in Pakistan: Model Conceptualisation, Configuration and Calibration.* Technical report. South Asia Sustainable Development Investment Portfolio (SDIP) Project. CSIRO.

Tsur, Y. (2009). On the economics of water allocation and pricing. *Annual Review of Resource Economics, 1*(1), 513–536.

United Nations Water (2015). *Eliminating Discrimination and Inequalities in Access to Water and Sanitation.* UN-Water Policy and Analytical Briefs.

Wang, M., Tang, T., Burek, P., Havlík, P., Krisztin, T., et al. (2019). Increasing nitrogen export to sea: a scenario analysis for the Indus River. *Science of the Total Environment, 694,* 133629.

Watto, M.A. (2015). The economics of groundwater irrigation in the Indus Basin, Pakistan: tube-well adoption, technical and irrigation water efficiency and optimal allocation. University of Western Australia, Doctoral thesis.

Watto, M.A., and Mugera, A.W. (2016). Irrigation water demand and implications for groundwater pricing in Pakistan. *Water Policy, 18*(3), 565–585.

Wheeler, S.A., Loch, A., Crase, L., Young, M., and Grafton, R.Q. (2017). Developing a water market readiness assessment framework. *Journal of Hydrology, 552,* 807–820.

WWAP (UNESCO World Water Assessment Programme) (2019). *The United Nations World Water Development Report 2019: Leaving No One Behind.* UNESCO.

WWF (2012). *Development of Integrated River Basin Management for Indus Basin: Challenges and Opportunities.* WWF, Islamabad. https://www.wwfpak.org/publication/pdf/irbm_2013.pdf.

Yang, Y.C.E., Brown, C.M., Yu, W.H., and Savitsky, A. (2013). An introduction to the IBMR, a hydro-economic model for climate change impact assessment in Pakistan's Indus River basin. *Water International, 38*(5), 632–650.

Yaqoob, A. (2016). Climate change and institutional capacity in the Indus Basin. *Spotlight on Regional Affairs, 35*(10). http://irs.org.pk/spotlight/spmoct16.pdf.

Young, W.J., Anwar, A., Bhatti, T., Borgomeo, E., Davies, S., et al. (2019). *Pakistan: Getting More from Water.* World Bank.

Yu, W., Yang, Y.-C., Savitsky, A., Alford, D., Brown, C., et al. (2013). *The Indus Basin of Pakistan: The Impacts of Climate Risks on Water and Agriculture.* World Bank.

Zhang, L., Wang, J., Huang, J., and S. Rozelle (2008). Development of groundwater markets in China: a glimpse into progress to date. *World Development, 36*(4), 706–726.

9. Water markets in France: appropriate water scarcity management mechanisms?

Simon de Bonviller and Arnaud de Bonviller

9.1 INTRODUCTION

In France, available renewable water resources per person exceed 2600 m³, while the overall withdrawal rate is around 19 per cent (Barthélémy and Verdier, 2008). France is therefore not a water-scarce country according to the Water Resources Vulnerability Index (Raskin et al., 1997) and water stress is local or occasional only (Falkenmark et al., 1989). Agriculture is the largest net consumer of water in the country, representing about 50 per cent of total net extractions (Roy, 2013). Irrigation in France developed significantly in the 1970s and 1980s, in a context influenced by the European Common Agricultural Policy subsidies (Martin, 2013) and favourable market conditions for irrigated crops. In the regions with the most irrigation-related withdrawals, water extractions often exceed the available renewable resource during low-flows months (Barthélémy and Verdier, 2008). The mismatch between water supply and demand frequently necessitates administrative bans and limitations during the summer (June to September).

In 2004, the European Water Framework Directive defined a good ecological state to be reached for all water bodies in the European Union (EU) member states. In order to reach this target, the 2006 French Water Law defined a new approach to quantitative water management, involving a shift from a private to a common property regime (Rinaudo, 2020). First, it required the definition of a maximum volume of water to be extracted (cap) in each basin. This volume (*volume prélevable*) was defined as the volume that can be fully extracted from the environment, on average eight years out of ten, all uses included, while ensuring the good functioning of the aquatic environment. Second, it required the revision of the existing water extraction authorizations in each basin, in order to comply with the cap. Third, it created Unique Organisms for Collective Management (OUGC) in areas where imbalance between water demand and supply occurred frequently (ZREs, for Zones de repartition des eaux).[1] OUGCs are responsible for irrigation water management and collectively request irrigation water rights.

The caps defined as a result of the 2006 Water Law often found a mismatch between irrigation water demand and allocations available, and several studies pointed towards significant economic losses if the corresponding cuts in water allocations were to be applied (Rouillard, 2020). This led to important delays in the implementation of the law (Martin, 2013), and to the formulation of compromises including the use of alternative water resources through the subsidized construction of substitution reservoirs.[2] In some contexts, such as the Adour–Garonne river basin (Luc, 2020), water markets have been mentioned as potential tools to mitigate such losses by redirecting water resources towards higher-value-added crops.

France is divided into seven river basins, each under the authority of an administrative Water Authority (Agences de l'Eau) established by the 1964 Water Law. Each basin undertakes a Water Development and Management Master Plan (Schémas Directeurs d'Aménagement et de Gestion des Eaux, or SDAGE) defined by Basin Committees and the establishment of Local Water Commissions (CLEs). Additional Local Water Resource Management Plans (Schéma d'Aménagement et de Gestion des Eaux, or SAGEs) can be established at the local level. The characteristics of each basin can vary widely (summary characteristics for different basins can be found in the Appendix, Table 9A.1). The seven French basins are relatively small (8700 to 155 000 km²) and highly populated (32 to 238 inhabitants per km²) in comparison to Australia's Murray–Darling Basin (1 059 000 km² with an average 1.9 per km² population density (ABS, 2008), where water markets have been extensively developed.

A debate on water markets in France was sparked by Strosser and Montginoul (2001), who provided a review of the economic principles underlying water markets and suggested two French contexts where the debate on water markets could be of interest: the Beauce aquifer and the Neste river basin. Barraqué (2002, 2004) reacted to this article by stressing the influence of common patrimony water management in France, as opposed to the use of privately owned water rights. Rinaudo et al. (2015) suggested the use of tradeable water savings certificates to improve urban water use efficiency. Different studies focused on the perception of water markets and economic incentives in general by stakeholders in France. These studies show a strong opposition to the use of markets applied to water resources on ethical grounds, with water seen as a 'common good' (Figureau et al., 2015). Fears of market power and third-party impacts, and an expected increase in financial resources needed for newcomers (Rinaudo et al., 2012), are other concerns, although informal trades occur (Kervarec, 2014). Qualitative analysis has shown that French farmers tend to prefer policies strengthening social incentives in relation to water over market mechanisms, although a contradiction with individualistic behaviours has been noted in practice (Rinaudo et al., 2016).

Some studies have attempted to model gains from trade arising from the potential use of water markets in France. Graveline and Mérel (2014) found that water markets could compensate 2 per cent of the economic losses generated by a 30 per cent reduction in water availability in the Beauce aquifer, equivalent to 3 cents per cubic metre. Potential explanations for this result include the fact that only 23 per cent of production was irrigated, and the possibility of using deficit irrigation without much yield loss on certain crops. These modest gains are consistent with findings from other French (Bouscasse and Duponteil, 2014), Spanish (e.g., Kahil et al., 2015) and Italian (Zavalloni et al., 2014) case studies integrated in the Water Cap and Trade European project (Rinaudo et al., 2014).

Water markets are not universal tools. In order for them to work in a given context, a range of enabling factors must be in place. Wheeler et al. (2017) provided a review of such factors, integrated in the water market readiness assessment (WMRA). This chapter applies the WMRA framework to two case studies: the Poitevin Marsh Basin, and the Neste river system. In order to document these applications, 11 semi-structured interviews were undertaken with key public, private and civil society stakeholders,[3] along with an extended literature review.

9.2   THE POITOU MARSH BASIN

9.2.1   Step 1A: Hydrology Considerations and System Type

The Poitou Marsh Basin (PMB) is a 6500 km² area located in western France, along the Atlantic Ocean coast. The Poitou Marsh itself (Marais Poitevin) represents 1000 km². Its land has been progressively recovered from the sea, from the 13th to the 20th century.

The Basin has a semi-circular shape and its external part supplies water to the Marsh. Mean temperatures are moderate (11°C) and rainfall is 850 mm/ year on average. Effective rainfall (about 280 mm) occurs between October and May. In particularly dry years, such as 2005, effective rainfall can be as limited as 50 mm with a total summer rainfall amounting to 100 mm (Douez et al., 2015).

Three main geological structures coexist from the heights to the shoreline, as shown by Figure 9.1. The primary bedrock is located in the Basin's northern part and gives birth to a dense surface water network. These rivers trickle down until they reach the second structure, formed by limestone and karst layers. These layers are porous, and the hydrographic network shows a lower density.

Many sources flow out of the karstic groundwater body, defining a part of the marsh called 'wet'. The wet marsh receives the water trickling down from the bedrock and the water from near groundwater bodies. The Poitou Marsh Basin ecompasses and supplies water to the Marsh through four main rivers, joining in two estuaries downstream.

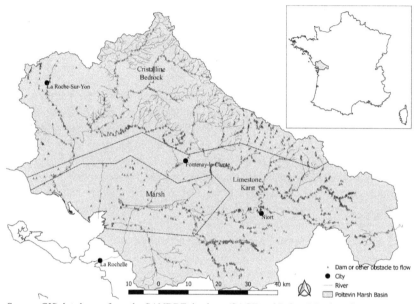

Source: GIS data layers from the SANDRE database (SAGE and Poitou Marsh perimeter, Dams), and the CARTHAGE database (rivers).

Figure 9.1   Poitou Marsh Basin geological structures

Water entering the Marsh flows through 8000 km of canals and ditches, whose water levels are monitored and supported by 200 dams spread throughout the territory. The canals are used to drain water from the Marsh during the winter and to store water in summer. The basin's hydrology has been evolving throughout centuries, since the Middle Ages when agriculture began in the basin. The hydrologic functioning of the Poitou Marsh basin is unusual, due to the existence of sources flowing out of the karstic aquifers, around the clay soils of the Marsh: in particular, the water supply of the Marsh can be deeply affected by a significant drop in the nearby interconnected aquifer water level.

The Poitou wetland hosts important ecological values linked to the Basin's different habitats: the 'wet' marsh, the 'dry' marsh and intermediary marshes are home to about 250 registered bird species, as well as a significant number of fish and vegetation species related to the marshes' physical, chemical and climatic characteristics (Ayphassorho et al., 2016). However, the wetland has been threatened in the last few decades by excessive water withdrawals, demographic pressure, water quality issues, the establishment of invasive species, and artificialization. France was condemned in 2000 by the European Court of Justice for failure to comply with the legislation on the protection of wild birds in the Poitou Marsh (EC, 1999).[4]

9.2.2   Step 1B: Existing Institutional, Planning and Property Rights Arrangements

Most (54.6 per cent) of the total water extractions within the Poitou Marsh basin are related to irrigation, followed by drinking water (42.7 per cent) and industrial water use (2.7 per cent) (EPMP, 2015). Irrigation water use occurs during the French summer between June and September, and spring (March to May) to a lesser extent. Irrigation withdrawals are concentrated within groundwater bodies in the dried marsh, located around (and upstream of) the wetland. As the groundwater and surface water bodies are largely interconnected throughout the Basin, excessive withdrawals in irrigation areas can lower the water level in the Marsh in summer, endangering environmental values and competing with other water uses (for example, drinking water, tourism, wastewater treatment).

Water governance in the Poitou Marsh Basin is characterized by a plurality of actors, and sometimes a lack of coordination (Ayphassorho et al., 2016). Three different formal instruments are currently used to manage water within the Basin. First, planification involves the definition of a Water Development and Management Master Plan (SDAGE) at the Loire–Bretagne Basin (upper) level, and three Local Water Resource Management Plans (SAGEs) at the local level. Second, state regulations have been designed to announce bans on irrigation water extractions in times of scarcity, and to enforce water rights. Third, different contracts have been established between various institutions (including Water Agencies) and local water users in order to reach the various goals defined by water management schemes.

Irrigation water rights in the Basin are collectively gathered and requested by a public state institution, the Etablissement Public du Marais Poitou (EPMP). The EPMP acts as a unique collective management organization (OUGC): it gathers all demands and collectively requests water rights to the state authorities, in conformity with the existing plans (SDAGE and SAGE). It then allocates water rights each year depending on the needs expressed by irrigators, in coordination with the local Agricultural Chambers. Water rights (*autorisations de prélèvements*) are granted on an annual basis and cannot be transferred, bought or sold.

During the irrigation season, excessive water extractions can drain water resources from the Marsh, endangering its ecological state. Therefore, crisis management measures have been decided based on the use of piezometric thresholds (minimum flows for surface water). If the water level in a ground-water body decreases under the first threshold (the alert threshold or *seuil d'alerte*), a temporary limitation of water quotas (10–40 per cent volume reductions) can be collectively applied by the EPMP on a voluntary basis. Under the second threshold (*seuil d'alerte renforcée*), the French state takes over and a legal 50 per cent cut of water quotas volumes is applied (*arrêté-cadre sécheresse*). Finally, the last threshold is defined as the ground-water level under which a ban of irrigation water use is legally declared, although some exceptions exist for particular high-valued crops.

The 2006 French Water Law required the definition of a cap (*volume prélev-able*) in all basins. As a result, a 2007 study (Groupe d'experts, 2007) provided a first estimation of the total extractions that could be ensured in eight years out of ten without compromising the aquatic environment in the Poitou Marsh Basin. As in many other contexts in France, the considered cap was consid-erably lower than current water extractions. This was later confirmed by the 2010 Leading Water Management Scheme (Comité de Bassin Loire-Bretagne, 2009), defining a 55 per cent reduction in water rights, and led to debates and fierce opposition from the agricultural sector expecting significant economic losses. As a result, no permanent cap was in place in the Poitou Marsh at the time of our study in 2019.

In order to reach a compromise, compensatory measures were decided by the state, including the construction of subsidized substitution reservoirs. Such reservoirs are filled during the winter, when the surface river flows and groundwater levels are high, and the stored water is used as a substitute to irri-gation water withdrawals between June and September. Such reservoirs were criticized by non-agricultural actors, pointing at high costs (€5 per cubic metre on average; Verley, 2020), potential environmental impacts and an explicit support to an intensive form of agriculture. In spite of legal actions filed by environmental non-governmental organizations (NGOs), in 2019 most of the substitution reservoirs were built or approved in the northern part of the Basin. This enabled a 30 per cent diminution of the annual volume of granted water rights in practice, although a significant part of it was related to reductions in unused water rights (Ayphassorho et al., 2016). In the southern part of the Basin, political controversies led to several mediations supervised by the state authorities. This led to a public management protocol specifying different agri-cultural practices to be adopted in exchange for the reservoir's construction and subsidies. Many projects and negotiations were still ongoing at the time of the study, and the associated reduction in water quotas was not finalized.

To summarize, in terms of the WMRA's Step 1, the 'Hydrology and system type' are well documented and monitored in the Poitou wetland, notably by the EPMP. An excess in water extractions endangering the Marsh in summer has been observed in the last few decades, suggesting the use of water scarcity management measures. Regarding the 'Institutional, planning and property rights arrangements', a collective approach has been encouraged by the 2006 legislation and the EPMP's action, and alternative approaches (that is, substitution reservoirs, collective restriction measures in times of scarcity) have been tested to manage water scarcity.

9.2.3   Step 2: Water Markets in the Poitou Marsh Basin: Potential Benefits, Impediments and Implementation

Step 2 of the WMRA focuses on the potential benefits from water trade. In order for such benefits to exist, there has to be a sufficiently diverse water demand. Irrigation represents most of the water demand in the Basin. The most important crops cultivated include corn (29 per cent of the total agricultural area) and maize (19 per cent). Other crops representing smaller agricultural areas include various other cereals, tobacco and seeds, and diverse higher-valued crops related to smaller water uses. Organic farming has been developing in the last few years, under what is generally perceived as favourable market circumstances. Drinking water is also responsible for a significant demand. Besides, tourism (the second-largest local economic activity, after agriculture) is deeply related to the Marsh's ecological values. Specifically, navigation in the emblematic parts of the Marsh represents an important water demand from the tourism sector. Thus, the demand for irrigation water is dominated by a few low-value crops (maize, corn …), while the demand for other water uses includes tourism and domestic water and involves a significant environmental water demand (EPMP, 2015).

9.2.4   Expected Benefits

In order to comply with the 2016–21 Water Development and Management Master Plan (SDAGE), significant cuts will be required to the annual volume of water rights granted (Comité de Bassin Loire-Bretagne, 2015). In this perspective, redirecting water from lower- to higher-value crops and less water-intensive crops would facilitate the transition while benefiting employment (Martin, 2013). The use of water markets could be considered to mitigate the economic losses related to a reduction in water extractions, and to incentivize such transfers. Bouscasse and Duponteil (2014) modelled irrigation water markets in the Poitou wetlands, considering a projected 55 per cent decrease in water quotas. They found that groundwater markets could mitigate only 0.5 per cent of the 18 per cent consecutive loss in gross margins, due to the relative homogeneity of farming systems, the hypothesized ability of farmers to adapt their cropping patterns using deficit irrigation, and the modelling approach (monthly demand and the use of a mean farming profile that leads to a lower modelled heterogeneity among farmers' profiles). They note, however, that achieving the same reduction in quotas using a pricing policy only (that is, increasing the water price without allowing water transfers) would cause a higher loss in farm gross margin.

The existence of important non-agricultural uses for water in the Basin suggests that with the existence of water markets, tourism actors and environmental water holders (that could involve the state or local public institutions) could buy water rights from the agricultural sector, thereby ensuring the strong functioning of aquatic environments and improving the agricultural sector's financial capacity to shift towards less water-intensive crops. The possibility of such trades is backed by anecdotal evidence, suggesting that a limited amount of informal water trade transactions might already be occurring in the Poitou Marsh Basin (Kervarec, 2014). Limited evidence of irrigators growing higher-valued crops leasing land to benefit from the associated water rights has also been reported in the interviews.

9.2.5 Impediments

Our analysis also revealed significant impediments to the use of water markets. The first important impediment is related to the French legislative framework. Currently, the annual water rights granted by OUGCs are not transferrable, and it is illegal to sell them. A second major impediment resides in the low social acceptability of water markets. Already shown by the scarce literature dedicated to water markets in France (Rinaudo et al., 2012, 2014, 2016; Figureau et al., 2015), this trend appears in the semi-structured interviews realized. While some interviewees considered that potential gains from trade did exist, they were worried about the redistributive effects of such policies. Water markets were alternatively perceived as endangering the ability to use water policy as a tool for climate change adaptation, favouring farmers financially at ease while hurting irrigators in a difficult financial situation, and representing a political choice hurting interviewees on ethical grounds. Concerns related to the concentration of water withdrawals near the Marsh and a resulting deteriorated public image of irrigators were also expressed. In this regard, we believe that establishing water markets in the Poitou Marsh Basin, or France in general, would require a deep paradigm change in the local attitudes towards water management.

The enforcement of existing water regulations was identified as another potential issue. Some water extractions occurring in the northern part of the basin (bedrock) were not monitored at the time of our study and were considered disconnected from the remaining hydrological system. Generally speaking, enforcement has been described as the 'Achilles heel' of French water policy (Montginoul et al., 2020). Strong hostility from the farming industry, lack of human resources, and political interference in times of water scarcity, can hinder the ability of the French water police to enforce the existing regulations and ensure that extractions match the existing water allocations. Nonetheless, in the Poitou Marsh Basin, the hydrological monitoring done by the EPMP and the reactive nature of the local hydrology (an increase in water use is reflected rapidly in groundwater levels) enable a reliable monitoring of local water resources availability and extractions.

As 70 per cent of the water extracted for irrigation purposes in the Poitou Marsh Basin is groundwater (EPMP, 2015) and most of the aquifers in the 'dry' marsh are connected, transportation costs would be low in the context of water transfers. Irrigation and drinking water use in the Basin can be compensated by the existing dams and (for the northern part) substitution reservoirs. Furthermore, the EPMP has established a public information system (SIEMP, available online) monitoring the water flows and levels for 204 stations across the Basin. Thus, environmental externalities related to water transfers could be adequately managed, provided that a cap is defined in compatibility with the environmental needs of the Marsh.

Such a cap, however, did not exist in the Poitou Marsh at the time of our study. Allowing trade without a cap would expose the wetland to additional pressures related to water withdrawals: although our interviews suggest that 'sleeping' water rights are rare in the Basin, the reduction in water extractions planned by the water management plan (SDAGE) has not yet been fully realized. Besides, groundwater and surface water resources are largely interconnected within and around the Marsh, and most water extractions occur in the Dogger groundwater aquifer, directly connected to the Marsh. Environmental externalities must thus be monitored closely.

Irrigation in the Poitou Marsh Basin has developed on an individual basis, without prior rules of water allocation. A second pattern of irrigation development can typically be found in France (Martin, 2013), where irrigation development has been fostered by a public–private institution, with collective water management rules defined prior to irrigation development. In order to cover an analysis of the second case, the following sections will apply the WMRA to the Neste system, in southwestern France.

9.3   THE NESTE SYSTEM

9.3.1   Step 1A: Hydrology Considerations and System Type

The Neste system (Figure 9.2) is located in southwestern France with a semi-oceanic climate. The average temperature is 13°C, and annual rainfall is around 750 mm. In the last few years, drought episodes have been common in the territory, with dry and hot summers. No significant groundwater aquifers can be found in the area.

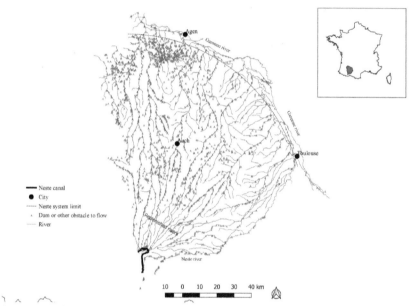

Sources: GIS data layers from the CARTHAGE and SANDRE databases.

Figure 9.2   The Neste system

In summer, most of the rivers forming the Neste system naturally dry up. In order to encourage the development of irrigation, the Neste canal was therefore built in 1962. This canal reconnects the rivers of the Neste system with the Neste river flowing from the Pyrenees mountains, alimented by snowmelt. By doing so, it replenishes 17 rivers through the use of channels totalling 90 km in length, and enables irrigation during the spring and summer seasons. The Neste canal has been historically operated by a public–private institution, the

Compagnie d'Aménagement des Coteaux de Gascogne (CACG). The canal and the 17 connected rivers are monitored in near real time by the CACG using a dedicated hydrological model.

Water scarcity in the Neste system is concentrated in the French summer (June to September), and can be experienced outside of irrigation withdrawal peaks in winter (December, January, February) when lower amounts of water flow from the Pyrenees.

9.3.2   Step 1B: Existing Institutional, Planning and Property Rights Arrangements

Water rights in the system are granted on an annual basis. Irrigators establish a contract with the CACG, which collectively requests water rights and delivers the water, in association with the local OUGC (the Gers Agricultural Chamber). An important feature of the Neste system is the *droit acquis* principle: once a water right is granted, a water user cannot be denied this right (in the same conditions) in the following years, as long as the user pays the related fee. As a result, water rights in the Neste system are bundled to land in practice, and the land value generally doubles if a water right is attached to it.

Unlike in the Poitou Marsh, a cap (*volume prélevable*) has been defined by the state authorities (the *préfet*) in 2012, as required by the 2006 Water Law. As water demand in the Neste system exceeds water supply, a waiting list has been created. Each year, priority is granted first to irrigators who request a right allocated to them in the previous year (*droit acquis*), second to newcomers, and third to irrigators wishing to increase their allocated volume.

Water scarcity management in the Neste system is based on thresholds, as in the Poitou Marsh. When the CACG detects that water flows fall below an alert threshold, the Neste Commission is convened. The Neste Commission involves a wide range of local water stakeholders. It can make temporary restrictions, and decisions are unanimity-based. Below a second threshold, water extractions are banned by a state regulation (*arrêté-cadre sécheresse*). Generally speaking, crisis management is facilitated by the fact that the Neste system is integrally monitored and flows can be adjusted by increasing or decreasing water extractions from the Neste River upstream.

9.3.3   Step 2: Water Markets in the Neste System: Potential Benefits, Impediments and Implementation

Irrigation represents the vast majority of water withdrawals from the Neste system. In 2009, annual irrigation water needs were estimated between 75 and 100 million cubic metres (Mm^3) depending on climatic circumstances. Drinking water and other domestic uses, in comparison, represented about

6 Mm³ of water withdrawals per year (Villocel et al., 2009), while industrial uses (including cooling of the Golfech nuclear power plant downstream of the system) were relatively low (about 2 Mm³).

About 70 per cent of the annual 220 Mm³ transiting from the Neste canal and through the Neste system is environmental water, meant to ensure that flows in the system remain over the defined minimum flow requirements (MFRs), ensuring services as dilution for water quality purposes. Other non-consumptive uses include navigation, fishing, hunting, diverse recreational uses such as kayaking, and hydropower (mills and dams). A convention has been established between hydropower use for the dams and reservoirs located upstream of the Neste system, where the hydropower operator can release water during low-flow periods but outside demand peaks for electricity, including a financial compensation for the loss incurred by the company (Fernandez et al., 2014). Specific minimum flows have been designed within the Neste system in this perspective.

The Neste system is a hydrologically controlled environment. Water use in the Neste system is therefore fully metered. Irrigators have strong incentives to use water as defined by their water right: in cases where water extractions exceed the water right, the CACG applies a tariff 4–11 times higher for the additional water used. The enforcement of water rights is therefore effective.

9.3.4 Expected Benefits from Water Markets

No study has been published on the potential economic benefits arising from the use of water markets and their related gains from trade in the Neste system. However, a potential benefit would be to fluidify the waiting list, allowing irrigators to buy water rights (Strosser and Montginoul, 2001). Note that the number of irrigators figuring in this list has been decreasing in recent years; additional research is ongoing to understand this evolution.

As extractions within the Neste system are compensated by water releases upstream and monitored in real time by the CACG, the potential environmental externalities arising from water transfers are limited. Although significant environmental flows are needed upstream in order to maintain the existing ecosystem, such externalities could be compensated by the flow control operated by the CACG, as long as the cap on total water extractions is enforced. The significant infrastructure in place would facilitate water transfer in the context of a water market, considerably limiting transaction costs.

9.3.5 Impediments

The Neste system shares impediments to the establishment of water markets with the Poitou Marsh Basin and with France in general: it is currently illegal to buy or sell water rights, and there is a strong cultural and political opposition to the use of markets to manage water resources (see the beginning of this chapter). In addition to this, water rights in the Neste system are bundled to land, due to the *droit acquis* principle: an irrigator owning a water right cannot currently be denied the same right in the next year. Abandoning this principle would expose farmers to a loss of prior investments, as the price of land increases when associated with an irrigation water right.

9.4 CONCLUSION

This chapter has applied the water market readiness assessment (Wheeler et al., 2017) to two case studies in France: the Poitou Marsh Basin and the Neste system. In terms of property rights/institutions, water management in France is not designed for water markets. It is illegal to buy, sell and (in most cases) transfer water extraction authorizations. In practice, water rights are still bundled to land, and the value of water rights is partly reflected in the increased value of land. In both our cases, the hydrology is well documented. A cap has been defined in most basins, including the Neste system. However, in the Poitou Marsh Basin it had not been implemented as of 2019. Uncertainties subsist regarding the impacts of climate change on future water availability, especially in the Poitou Marshes. In both cases, trade impacts are not well understood, as no formal trade has been occurring until now.

Externalities are not expected to be a concern in the Neste system, as it is under constant monitoring through the Neste canal. In the Poitou Marsh Basin, a well-known potential externality is that excessive withdrawals in the Dogger aquifer can drain water from the Marsh in summer. In terms of governance, monitoring and enforcement are effective in the Neste system. However, in the Poitou Marsh Basin it is less clear.

System types in our case studies involve resources that are suitable for trade, be it irrigation groundwater in the Poitou Marsh Basin or surface irrigation water in the Neste system. Both systems include ditches and canals that would enable water transfers in practice. However, trading water in the Poitou Marsh Basin would require significant regulation requirements to avoid negative externalities affecting the Marsh. Within the Poitou Marsh Basin, important reductions in the volume of water quotas granted annually will be required in the coming years, due to climate change and environmental constraints. Water markets could be considered as a tool to limit the resulting economic losses in the agricultural sector. Transfers involving environmental water holders (that could involve the state, civil society and/or the tourism sector) could be considered to protect environmental values in the Marsh. In the Neste system, as climate change is expected to further reduce average and minimum flows in the area (CBAG, 2017), additional means of demand management and flexibility would be of interest.

However, significant impediments have been identified to the use of water markets in our case studies. In the Neste system, implementing water markets would require unbundling water rights from land, thereby exposing irrigators to important losses in capital investment. The most important impediment is the very low social acceptability of water markets in France. Concerns and opposition to the use of water market mechanisms applied to water resources (redistributive effects, ethics and water as a common good, fear of monopoly power ...) have been expressed across our interviews and clearly described by the literature. Water in France is considered by the 1992 Water Law as a 'common patrimony' of the nation, and the notion of common patrimony has often influenced the French water management framework (Calvo-Mendieta et al., 2017). In this perspective, it seems that implementing water markets in France would require considerable change in the local paradigms of water management, in a context where existing frameworks are already subjected to significant political debates.

NOTES

1. A map depicting all ZREs can be found in the Appendix, Figure 9A.1.
2. Substitution reservoirs are filled in winter (when more water is available) and used in times of scarcity, as a substitute for groundwater or surface water pumping.
3. Ten interviews were organized face to face and one by telephone, involving 13 persons in total. Five interviews were dedicated to the Neste system, and six to the Poitevin Marsh Basin. Interviews lasted between one and two hours, and mainly focused on questions guiding the assessment described in Wheeler et al. (2017), while some also mentioned topics deemed important and/or interesting by the interviewee(s). A descriptive summary of all interviews can be found in the Appendix, Table 9A.2.
4. https://curia.europa.eu/en/actu/communiques/cp99/aff/cp9993en.htm, accessed 29/04/2019.

REFERENCES

ABS (2008). *Water and the Murray–Darling Basin – A Statistical Profile, 2000–01 to 2005–06*, 4610.0.55.007. Canberra, Australian Bureau of Statistics.
Ayphassorho, H., Caude, G., and Etaix, C. (2016). *Le Marais Poitevin: État Des Lieux Actualisé Des Actions Menées à La Suite Du Plan Gouvernemental 2003-2013 et Orientations, 005928–05*. Ministère de l'Environnement, de l'Energie et de la Mer.
Barraqué, B. (2002). Les marchés de l'eau en Californie: modèle pour le monde, ou spécificité de l'ouest aride américain? Première partie: la crise du partage du Colorado. *Annales Des Mines*, 28, 71–82.

Barraqué, B. (2004). Les marchés de l'eau en Californie: modèle pour le monde ou spécificité de l'ouest aride américain? Deuxième partie: marchés de l'eau ou économies d'eau? *Annales Des Mines*, *33*, 60–68.

Barthélémy, N., and Verdier, L. (2008). *Les Marchés de Quotas Dans La Gestion de l'eau: Les Exemples de l'Australie et de La Californie*. Paris, Commissariat Général au Développement Durable.

Bouscasse, H., and Duponteil, A. (2014). Agricultural water markets in Marais Poitevin. In J.D. Rinaudo (ed.), *Water Market Scenarios for Southern Europe: New Solutions for Coping with Increasing Water Scarcity and Risk?* (pp. 32–33). IWRM-NET.

Calvo-Mendieta, I., Petit, O., and Vivien, F.-D. (2017). Common patrimony: a concept to analyze collective natural resource management. the case of water management in France. *Ecological Economics*, *137*, 126–132.

Comité de Bassin Adour-Garonne (CBAG) (2017). *Changements Climatiques En Adour-Garonne: Notre Avenir Passe Par l'eau*.

Comité de Bassin Loire-Bretagne (2009). *Schéma Directeur d'aménagement et de Gestion Des Eaux (SDAGE) Du Bassin Loire-Bretagne 2010–2015*.

Comité de Bassin Loire-Bretagne (2015). *Schéma Directeur d'aménagement et de Gestion Des Eaux 2016–2021, Bassin Loire-Bretagne*.

Douez, O., Bichot, F., and Dupeuty, J.-E. (2015). *Etablissement Public Du Marais Poitevin: Etude d'impact Pour l'obtention de l'autorisation Unique de Prélèvement*. EPMP.

EPMP (2015). *Atlas Du Marais Poitevin*.

European Commission (EC) (1999). Press Release NO 93/99: Failure by France to comply with the legislation on the protection of wild birds in the Poitevin Marsh results in another ruling against it by the Court of Justice. https://curia.europa.eu/en/actu/communiques/cp99/aff/cp9993en.htm.

Falkenmark, M., Lundqvist, J., and Widstrand, C. (1989). Macro-scale water scarcity requires micro-scale approaches. *Natural Resources Forum*, *13*(4), 258–267.

Fernandez, S., Bouleau, G., and Treyer, S. (2014). Bringing politics back into water planning scenarios in Europe. *Journal of Hydrology*, *518*, 17–27.

Figureau, A.-G., Montginoul, M., and Rinaudo, J.D. (2015). Policy instruments for decentralized management of agricultural groundwater abstraction: a participatory evaluation. *Ecological Economics*, *119*, 147–157.

Graveline, N., and Mérel, P. (2014). Intensive and extensive margin adjustments to water scarcity in France's Cereal Belt. *European Review of Agricultural Economics*, *41*(5), 707–743.

Groupe d'experts (2007). *Rapport du groupe d'experts mis en place à la demande du ministère chargé de l'écologie sur les niveaux d'eau dans le Marais Poitevin, la piézométrie des nappes de bordure et les volumes prélevables pour l'irrigation dans le périmètre des SAGE du Lay, de la Vendée, de la Sèvre-Niortaise et du Marais Poitevin*.

Kahil, M.T., Dinar, A., and Albiac, J. (2015). Modelling water scarcity and droughts for policy adaptation to climate change in arid and semi-arid regions. *Journal of Hydrology*, *522*, 95–109.

Kervarec, F. (2014). Perception of water markets in Marais Poitevin (France). In J.D. Rinaudo (ed.), *Water Market Scenarios for Southern Europe: New Solutions for Coping with Increasing Water Scarcity and Risk?* IWRM-NET.

Luc, A. (2020). Conceptual approaches, methods and models used to assess abstraction limits for unconfined aquifers in France. In J.D. Rinaudo, C. Holley, S. Barnett and M. Montginoul (eds), *Sustainable Groundwater Management* (pp. 211–227). Springer.

Martin, P. (2013). *La Gestion Quantitative de l'eau En Agriculture. Une Nouvelle Vision, Pour Un Meilleur Partage.* Parliamentary Report to the Prime Minister, Paris, Assemblée Nationale.

Montginoul, M., Rinaudo, J.D., and Alcouffe, C. (2020). Compliance and enforcement: the Achilles heel of French water policy. In J.D. Rinaudo, C. Holley, S. Barnett and M. Montginoul (eds), *Sustainable Groundwater Management* (pp. 435–459). Springer.

Raskin, P., Gleick, P., Kirshen, P., Pontius, G., and Strzepeck, K. (1997). *Water Futures: Assessment of Long-Range Patterns and Problems.* Stockholm Environment Institute.

Rinaudo, J.D. (2020). Groundwater policy in France: from private to collective management. In J.D. Rinaudo, C. Holley, S. Barnett and M. Montginoul (eds), *Sustainable Groundwater Management* (pp. 47–65). Springer.

Rinaudo, J.D., Berbel, J., Calatrava, J., Duponteil, A., Giannocarro, G., et al. (2014). *Water Market Scenarios for Southern Europe: New Solutions for Coping with Increased Scarcity and Drought Risk?* IWRM-NET.

Rinaudo, J.D., Calatrava, J., and Vernier de Byans, M. (2015). Tradable water saving certificates to improve urban water use efficiency: an ex-ante evaluation in a French case study. *Australian Journal of Agricultural and Resource Economics*, *60*(3), 422–441.

Rinaudo, J.D., Montginoul, M., Varanda, M., and Bento, S. (2012). Envisioning innovative groundwater regulation policies through scenario workshops in France and Portugal. *Irrigation and Drainage*, *61*(1), 65–74.

Rinaudo, J.D., Moreau, C., and Garin, P. (2016). Social justice and groundwater allocation in agriculture: a French case study. In A.J. Jakeman, O. Barreteau, R.J. Hunt, J.-D. Rinaudo and A. Ross (eds), *Integrated Groundwater Management* (pp. 273–293). Springer.

Rouillard, J. (2020). Tracing the impact of agricultural policies on irrigation water demand and groundwater extraction in France. In J.D. Rinaudo, C. Holley, S. Barnett and M. Montginoul (eds), *Sustainable Groundwater Management* (pp. 461–479). Springer.

Roy, L. (2013). Gestion quantitative de l'eau et irrigation en France. *Sciences Eaux & Territoires*, *11*(2), 4–5.

Strosser, P., and Montginoul, M. (2001). Vers des marchés de l'eau en France? Quelques éléments de reflexion. *Annales Des Mines*, *23*, 13–31.

Verley, F. (2020). Lessons from twenty years of local volumetric groundwater management: the case of the Beauce aquifer, central France. In J.D. Rinaudo, C. Holley, S. Barnett and M. Montginoul (eds), *Sustainable Groundwater Management* (pp. 93–110). Springer.

Villocel, A., Boubee, D., Lagardelle, G., and Chisné, P. (2009) *Soutien Des Étiages Dans Le Sud-Ouest de La France: Outils de Gestion Équilibrée de La Ressource En Eau.* Lyon.

Wheeler, S.A., Loch, A., Crase, L., Young, M., and Grafton, R.Q. (2017). Developing a water market readiness assessment framework. *Journal of Hydrology*, *552*, 807–820.

Zavalloni, M., Raggi, M., and Viaggi, D. (2014). Water harvesting reservoirs with internal water reallocation: a case study in Emilia Romagna, Italy. *Journal of Water Supply: Research and Technology – Aqua*, *63*(6), 489–496.

APPENDIX

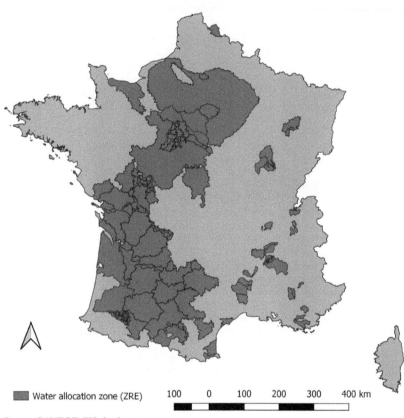

Water allocation zone (ZRE)

100   0   100   200   300   400 km

Source: SANDRE GIS database.

Figure 9A.1   Zones of Water Allocation (ZREs)

Table 9A.1   *The seven French basins*

Basin	Size (km²)	Main river	Mean population density (hab/km²)
Loire–Bretagne	155 000	Loire	83.0
Seine–Normandie	95 000	Seine	192.6
Rhin–Meuse	31 400	Rhin	136.9
Artois–Picardie	20 000	Escaut	238.0
Adour–Garonne	117 650	Garonne	59.5
Rhône Méditerranée	130 000	Rhône	116.0
Corse	8700	-	32.0

Source: French basin authorities.

Table 9A.2   *Semi-structured interviews undertaken*

Type of actor	Institution or company	Interviews	Interviewees involved
State representatives	Etablissement public du Marais Poitevin (EPMP)	1	2
Local government institutions	Syndicat mixte du Lay	1	1
	Syndicat mixte Vendée Sèvre Autizes	1	1
	Institution Interdépartementale du Bassin de la Sèvre Niortaise (IIBSN)	1	1
Civil society representatives	Chambre d'Agriculture des Deux-Sèvres	1	2
	Chambre d'Agriculture du Gers	1	1
	France Nature Environnement (FNE) Deux-Sèvres	1	1
	Fédération de pêche Midi-Pyrénées	1	1
Private sector	Compagnie d'Aménagement des Coteaux de Gascogne (CACG)	3	3

10. Best-laid plans: water markets in Italy
C. Dionisio Pérez-Blanco

10.1 INTRODUCTION

In Italy, in 2011, 95.8 per cent of the participants in a nationwide referendum voted against compensating water suppliers beyond the cost recovery of capital investments. This popular mandate revoked a 2006 legislative decree that made water suppliers eligible to exact profit from water provision services, effectively negating the possibility to achieve Kaldor–Hicks efficiency improvements through water trading (GU, 2011). Since then, the development of water markets has remained stagnant, and water policy focus has shifted to reactive drought responses based on voluntary agreements and/or command-and-control approaches that strengthen water allocation constraints (Pérez-Blanco et al., 2017). Water markets in Italy are thus situated in Step One of the water markets readiness assessment framework (WMRA): background context (Wheeler et al., 2017); with little progress having been made towards Step Two of the WMRA: net benefits of trade . Accordingly, this chapter focuses on the background context for the development of water markets in Italy (that is, Step One of the WMRA), providing an analysis of: (1) hydrology considerations and system type (in section 10.2); and (2) existing institutional and planning and property right arrangements (section 10.3). Finally, the chapter explores some critical issues for the transition towards Step Two of the WMRA through an assessment of current gaps or requirements for reform (section 10.4).

In this analysis, examples are provided from the Po River Basin District (PRBD) in Northern Italy (Figure 10.1), the largest and most economically relevant river basin in Italy. The reason for focusing on a specific river basin rather than conducting a nationwide analysis lies in the complex and decentralized planning and property rights and institutional arrangements in place in Italy. Italian regions have jurisdiction over the Water Abstraction Licence (WAL), issuing licences and setting water tariffs to be paid by users, which typically result in different WAL regimes coexisting within the same river basin system. The WAL census is incomplete, and information on actual withdrawals (that is, water removed from surface or groundwater bodies for any use), let alone water consumption (that is, water that is converted to vapour via evaporation and transpiration and becomes unavailable for further local use), is often unavailable. Despite a recent push towards homogenization through the Permanent Observatory of Water Uses (AdBPo, 2016a), an institutional structure that monitors water availability trends and coordinates responses to droughts through voluntary water reallocation agreements, institutional arrangements to manage water scarcity and droughts remain largely heterogeneous. This makes a comprehensive assessment of the WAL regime and institutional setting of Italian basins a daunting task, which cannot be addressed in a single chapter.

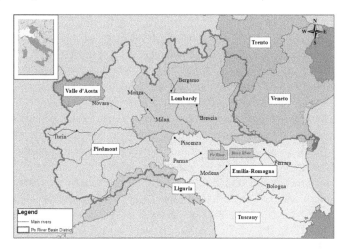

Figure 10.1   The Po River Basin District in Northern Italy

With an extent of 71000 km², the PRBD represents 24 per cent of the Italian territory, is home to 30 per cent of the Italian population and produces around 35 per cent of Italy's gross domestic product (GDP). It encompasses the territory of the Italian regions of Valle d'Aosta, Piedmont, Lombardy and Emilia-Romagna (including the large Italian cities of Milan, Turin, Bologna, Brescia, Modena, Parma Reggio Emilia, Ferrara, Monza, Bergamo, Novara and Piacenza); and a marginal fraction of the regions of Veneto, Liguria and Tuscany. Agriculture is the largest consumptive water use, representing 80.3 per cent of total withdrawals in the Po River Basin District , followed by households (12.2 per cent) and industry (7.5 per cent). In addition, the PRBD hosts 41 per cent of Italy's hydropower potential and 50 per cent of Italy's thermo-electric energy production, which results in significant non-consumptive water demands (AdBPo, 2016b).

Application of the WMRA to the PRBD suggests that Italy is still distant from the implementation of a full-fledged water market. Italy faces a challenging range of future climate and hydrologic scenarios, an under-developed and intricate regulatory architecture, and public opposition to trading as revealed through the 2011 referendum results. It is suggested that in order to progress towards the implementation of water markets, Italy will need to deploy regulatory and institutional arrangements that avoid over-allocation, prevent externalities and limit third-party impacts, so to enable efficient water markets that are also socially and politically acceptable.

10.2   A SHIFTING HYDROLOGICAL REGIME

The first stage in the implementation of the WMRA – background context – 'is a scoping exercise that establishes the context of a proposed market' (Wheeler et al., 2017), and includes the analysis of: (1) hydrology considerations and system type; and (2) existing institutional and planning and property right arrangements . This section focuses on the former.

The European Environment Agency (EEA) periodically develops a set of 46 climatic indicators measuring medium- to long-term climate change impacts (EEA, 2018, 2019). Under the Representative Concentration Pathway (RCP) 8.5 scenario, results for Italy show a reduction in average rainfall , along with more intense and less frequent extreme precipitation events; a reduction in snow cover; a reduction in runoff (up to –40 per cent); rising water temperatures (circa +0.1–0.3°C per decade in the lakes of Northern Italy); reduced soil humidity; and up to 20 per cent average increase in agricultural water withdrawals. Although EEA indicators typically report similar trends for the country (EEA, 2018), the magnitude of the impacts can differ considerably between macro-regions, and with significant differences between, for example, Northern Italy and Southern Italy and the islands.

A growing number of downscaling studies assess climate change impacts on water resources in Northern Italy and the Po River Basin in particular (see Table 10.1). Available studies reveal fluctuations in precipitation, river regime, and flooding and drought frequency. Over the last 60–80 years, precipitation intensity per rainfall event has shown a growing trend, while annual precipitation has decreased (Brunetti et al., 2000; Forzieri et al., 2014). Glacier retreat and permafrost reduction will potentially create large Alpine lakes that increase flood risk (Italian Ministry of Environment, Land and Sea, 2013). The spring discharge peak will shift from May to April due to changes in snowmelt timing, reducing runoff from May to November when water demand is at its maximum (Angelini and Reiterer, 2014; Coppola et al., 2014; Groppelli et al., 2011).

Table 10.1   *Climate change impacts on the water system in the PRBD,*
                Northern Italy

Area	Results	Source
Northern Italy	Using daily precipitation data over five meteorological stations in Genoa (1833–1998), Milan (1858–1998), Mantova (1868–1997), Bologna (1879–1998) and Ferrara (1879–1996), the authors test the hypothesis of growing precipitation intensity in Northern Italy. It is found that 'the number of rainy days has a stronger and more significant negative trend than total rainfall', suggesting that precipitation intensity has a positive trend in Northern Italy. This is particularly evident during the last 60–80 years of the series. These results show an increase of the total rainfall, as well as of heavy precipitation events of 25 mm and 50 mm per day.	Brunetti et al., 2000
Italy – focus on Po River Basin	The authors use 12 climate experiments derived from a combination of four global circulation models and seven regional climate models to force the geographic information system (GIS)-based hydrological model LISFLOOD. Data series used for validation correspond only to Northern Italy along the Po Valley. Water withdrawals and consumptions in different sectors are also assessed using WaterGAP3, a platform consisting of a set of sub-models. Intense water abstractions in Italy (>100 mm) will increase by 25 per cent in the north of the country due to intensification of crop production, and slightly decrease in the south. Minimum flows in Italy are expected to be reduced by up to 40 per cent by the 2080s. In Northern Italy 'the reductions in minimum flows are relatively more severe for smaller return periods'. Twenty-year return period event deficit volumes will increase by 50 per cent by 2020 and up to 80 per cent by the 2080s. In Alpine regions, streamflow deficits are expected to increase between 20 and 50 per cent from 2050s onwards.	Forzieri et al., 2014
Alps – Oglio Alpine Watershed	The authors 'downscale future precipitation and temperature from three General Circulation Models (GCMs) and use this data to feed a minimal hydrological model to investigate future hydrological cycle (2045–2054)'. Projected precipitation and temperature differ depending on the GCM adopted, and so do the hydrological scenarios. 'A "relatively good agreement" is found upon expected thinner and shorter seasonal snow cover and increased evapotranspiration'. Yet, precipitation projections are less consistent.	Groppelli et al., 2011
Alps	Temperature increase recorded in the Alps during the last 30 years is threefold the average of the entire Northern Hemisphere (+1°C in the annual mean, more pronounced in summer). Snow height series during 1920–2005 in 41 stations reveal a decreasing trend, especially in the last 30 years (–18 per cent as compared to 1959–2002, and up to –40 per cent in some areas). In future years, average temperature will increase from 2°C to 6°C, rainfall will suffer marked seasonal variations leading to more frequent and intense extreme events, minimum snow presence altitude will increase, and glaciers will retreat in mass. All this will lead to a reduction in summer runoff, an increase in winter runoff, and higher landslide and glacial risk.	Angelini and Reiterer, 2014

Area	Results	Source
Italy – focus on the Alps	In coastal freshwater beds, saltwater intrusion and loss of wetlands are expected. In the Alps, increasing intensity and frequency of precipitation events in winter and decreasing in summer, glaciers retreating and permafrost reduction, are all expected; which may cause the potential formation of large Alpine lakes as glaciers retreat, which might cause glacier lake outburst floods. As compared to the 1971–90 average, small glaciers are expected to disappear, larger glaciers will suffer a reduction between 30 and 70 per cent of their volume by 2050, and the ice cover will be reduced by 80 per cent by 2050. In Northern Italy, more intense precipitation and floods are expected. In Southern Italy, precipitation decrease and summer droughts are projected.	Italian Ministry of Environment, Land and Sea, 2013
Italy	Primary concerns relate to water quantity, extreme weather events (droughts in southern regions; hydrogeological risks in the Alpine area) and resultant loss of biodiversity. Water stress may increase by 25 per cent in this century as a result of growing irrigation demand. Ensuring safe water supply will be especially challenging in southern regions and the islands (Apulia, Basilicata, Sardinia, Sicily). Reduction of water availability will be also felt in Central and Northern Italy, especially along the Po River Basin. Saltwater intrusion will deplete water quality and stocks in coastal freshwater aquifers and wetlands. Alpine water regime will be affected through precipitation variations, reduced snow cover, glacier retreat and reduced glacier storage. This will result in increased winter runoff (+90 per cent), reduced summer runoff (–45 per cent), more summer droughts and floods and landslides in winter, including floods 'due to glacial lake outburst'. Flood events will also increase in the south.	OECD, 2013
Alps (Piedmont) – Toce Watershed	The study assessed climate change impacts on hydrological processes using 'two models: i) a simplified hydrological model that uses only precipitation and temperature to compute the hydrological balance; and ii) an enhanced version of the model that computes the actual evapotranspiration using an energy balance equation'. Two climate models (REMO and RegCM3) are used to force the hydrological models. Mean monthly discharge for the period 2041–50 as compared to 2001–10 increases between 36 and 68 per cent in January, between 81 and 119 per cent in February, between 48 and 126 per cent in October, and between 9 and 49 per cent in December. There is also a projected decrease in the summer months (maximum of –36 per cent in August). Importantly, the simplified model yields similar results, suggesting that simple hydrological models can deliver satisfactory results in Alpine basins where 'meteorological forcings with the high space and time resolutions required to run more sophisticated models' may not be available.	Ravazzani et al., 2014

Area	Results	Source
Upper Po River Basin	Two climate models (REMO and RegCM) are used to force an ensemble of eight hydrological simulations under the A1B emission scenario. This research compares periods 2020–50 and 1960–90. Results show a displacement of the spring peak in discharge flows from May to April, entirely explained by the change in snowmelt timing. Runoff decreases through the entire year with the exception of winter. Winter runoff increase is concentrated in the northern area (+20 per cent in low elevation areas and +40 per cent in high elevation areas). In the autumn, discharge drops by 20 per cent everywhere, and up to –40 per cent in some areas to the far north and south of the basin. During spring, discharge drops up to 20 per cent.	Coppola et al., 2014
Po River Basin	Precipitation and temperature 'under RCP4.5 and RCP8.5 are used to obtain climate projections nesting the regional climate model COSMO-CLM into the global climate model CMCC-CM'. Outputs are used to force the hydrological model and compare impacts on hydrological processes for the 2041–70 and 2071–2100 periods versus 1982–2011. Average discharges decrease under both RCPs. For the 2041–70 period, discharges fall from May to November and remain constant the rest of the year. For 2071–2100 projections, 'discharge reduction from May to November persists and is more severe than in the preceding period', and increases up to 60 per cent in the rest of the year.	Vezzoli et al., 2015
Po River Basin	Using graphical and analytical tools the authors assess the variability of river discharge to distinguish human-induced from natural perturbations on flows. The analysis reveals fluctuations in the river regime and flooding and drought frequency that seem to stem from perturbations 'whose memory is maintained in the long term', suggesting permanent rather than temporary changes in observed trends that may be attributable to human-induced climate change.	Montanari, 2012
Po River Basin	The authors extend available discharge series by 110 years, collecting data from the historical archives of the Hydrological Office of the Po River Basin Authority. This data is complemented with data on evapotranspiration and precipitation for the period 1831–2003. The authors observe a regime shift since the 1920s (progressive reservoir depletion) that coincides with a downward shift in precipitation and an upward shift in evapotranspiration. The increase in peak flow discharges is 'apparently the result of levee works', that is, they cannot be attributed to climate change based on the paper's findings.	Zanchettin et al., 2008

The climatic variations in the coming decades will influence the agricultural sector and its production dynamics, especially in vulnerable areas along the Mediterranean coast. The agricultural production system will be subject to variations in terms of duration of the phenological/plant life cycle, productivity and displacement of typical cultivation areas (northward and towards higher altitudes). The intensity and sign of the changes will vary depending on the plant species considered (OECD, 2013). Overall, higher temperatures will reduce the length of the growing cycle, leading to lower biomass accumulation and a consequential reduction in yields. Greatest yield reductions are foreseen for spring–summer cycle crops (maize, sunflower, soy). Crops such as wheat, rice and barley may partially offset the negative impacts of the changed climatic conditions as they are able to respond more efficiently to the direct effects of the increase in atmospheric carbon dioxide (CO_2) concentration than for example maize, sorghum and millet (IPCC, 2018). Among other adaptive responses, irrigators are expected to increase their water demand. As a result, intense water abstractions (>100 mm) will increase by 25 per cent in the north of the country (Forzieri et al., 2014).

The combined effects of growing demand and diminishing supply are expected to reduce minimum discharge flows by up to 40 per cent by the 2080s and significantly limit groundwater recharge in Northern Italy (Forzieri et al., 2014). Evidence of some of these changes is already noticeable: average temperature in the basin has increased by around 1°C since 2000 as compared to the historical data series; average rainfall records show a downward trend; and emergency droughts (discharges below 500 m³/s in the Po River outlet) have hit the basin with increasing frequency and intensity after the turn of the century: four times in 19 years (2003, 2004, 2007 and 2017), as compared to seven times in the previous century (1938, 1941–42, 1962, 1965, 1970, 1973–74 and 1988–90) (AdBPo, 2010; Pérez-Blanco et al., 2017; Tibaldi et al., 2010). These trends will make the current WAL regime unsustainable under current use arrangements, suggesting that reallocation instruments will be necessary to ensure sustainable water use and mitigate economic losses (Santato et al., 2016).

A National Climate Change Adaptation Strategy (Strategia Nazionale di Adattamento ai Cambiamenti Climatici, SNACC) was approved in 2015, and further developed into a National Adaptation Plan for Climate Change (Piano Nazionale di Adattamento ai Cambiamenti Climatici, PNACC) in 2017 (MATTM, 2017). The PNACC defines climate change scenarios for Italy, gathers evidence on their potential impacts and encourages relevant institutions (river basin authorities, in the case of water resources management) to identify and adopt effective adaptation strategies to climate change, including 'water reallocation instruments' such as markets (MATTM, 2017).

10.3 THE WATER ABSTRACTION LICENCE (WAL) REGIME

The Italian WAL regime creates a significant obstacle to the implementation of water reallocation instruments such as trade. Italian laws have traditionally regarded water as a plentiful resource, which has led to *de jure* water resources that are over-allocated and subject to limited monitoring and enforcement.

During the 20th century, the Italian WAL regime has undergone a process of decentralization whereby regions have obtained full control over WAL administration, with limited to no oversight from basin authorities or the central government. This has resulted in poor coordination and over-allocation. At present, hydropower and agricultural surface WALs in the PRBD amount to 1840 m³/s, which is about 25 per cent larger than the average river flow of 1470 m³/s (MATTM, 2014, 2017). Only the large volume of 'sleeper' licences prevents the effective dewatering of the Po River and its tributaries; a situation that is likely to change due to global warming and growing irrigation demand trends.

Moreover, the PRBD and other Italian basins lack a central WAL register: each region collects its own information on existing WALs, which results in multiple, asymmetric and often incomplete and non-comparable data. A recent study by Santato et al. (2016, p. 9) found that regional censuses in the PRBD typically report WALs 'in absolute terms, either as average or (less frequently) maximum volume of flow that can be withdrawn'. A large fraction of WALs, particularly in agriculture, do not specify the abstraction volume. This is aggravated by an insufficient cost-recovery process that does not reflect the costs of water storage, delivery or use, thus giving users the possibility to expand water withdrawals at relatively low cost where needed (for example, following the development of new irrigable land, which will be exacerbated due to climate change adaptive responses; Forzieri et al., 2014). Although the duration of WAL entitlements is limited (between 15 and 40 years in the PRBD), they are renewed automatically provided use continues, and public intervention to restore the balance is likely to incur large, potentially disproportionate institutional transaction costs (Rey et al., 2018). This 'tortuous and substandard' WAL 'hinders the performance of bottom-up conflict resolution mechanisms at a basin scale', such as markets (Santato et al., 2016, p. 2).

The transposition of the European Union (EU) Water Framework Directive (OJ, 2000) in Italy resulted in the consolidation of concepts, definitions and laws that acknowledge nature's limits and articulate the economic uses of water with respect to environmental protection and integrated water resources management; although this is not substantiated yet in regional WAL regimes (Santato et al., 2016). This dichotomy has been pointed at as a major barrier towards sustainable water use and economic development, with both the SNACC and the PNACC calling for a revision of the WAL regime (MATTM, 2014, 2017). In the early 2000s the government sought to intervene and reorganize the WAL regime. The most notable reform happened in 2006 and involved what was then seen as a first move towards the introduction of water trading in Italy. In its article 154, the Legislative Decree 152/2006 read: 'the water tariff ... shall be determined taking into account the quality of the water resource and service provided, the investments in water works, the costs of managing water works, the adequate remuneration of the capital invested and the management costs of protected areas' (GU, 2006, p. 1). From 2008 to 2010, the Italian government also approved a series of reforms that handed over the management of economically relevant local public services (including water supply and sanitation services) to private agents or public companies managed by private agents (at least 40 per cent share), through contracts awarded via competitive public tenders, allowing in-house (public) management only when very exceptional situations impeding an effective recourse to the market arose (GU, 2008a, 2008b, 2009a, 2009b, 2009c, 2010). This shift towards privatization of water supply and sanitation services was presented as a means to improve conveyance efficiency in the water pipe network system, which was then at 70 per cent on average. Yet, the Italian population perceived these legal developments as an attempt to exact private revenues from water resource management, which could eventually limit access to a fundamental right, echoing views that were becoming widespread in Europe at the time through citizens' movements such as Right2Water.

In a referendum in 2011, a landslide vote repealed the privatization of water supply and sanitation services (95.4 per cent of the votes) and the possibility to compensate water suppliers beyond cost recovery (95.8 per cent), effectively halting progress towards the implementation of water markets in Italy (GU, 2011). These results, in particular the partial abrogation of article 154 of the Legislative Decree 152/2006, essentially prevent the adoption of formal market reallocation instruments such as trading or buyback; and explain the shifting focus of Italian institutions towards non-pecuniary reallocation arrangements (for example, voluntary agreements through the Permanent Drought Observatory) and volumetric water charging instruments that allow for more accurate, higher cost recovery ratios (Pineschi and Colaizzi, 2014; Regione Piemonte, 2017) as a means to reallocate water resources among uses.

10.4   DISCUSSION AND CONCLUDING REMARKS: GAPS AND REQUIREMENTS FOR REFORM

Following the 2011 referendum results, the prevailing view in Italy is that access to water is a fundamental right that should not be subject to market reallocation (Servizio Studi – Dipartimento ambiente della Camera dei Deputati, 2011). This is not to say that markets have vanished from the policy debate: researchers and policy-makers in Italy and the EU acknowledge the potential role of markets as a tool to reallocate water from low- to higher-value economic uses; and as a realpolitik instrument to ensure reserves for the public good where institutional lock-in prevents compliance with environmental codes and laws, for example, through buyback (Gómez et al., 2017). The challenge with water markets in Italy today is that of how to enhance economic efficiency in water resources allocation without compromising water resources available for the public good (household and environmental uses), so as to increase acceptability among the wider public.

In developing a water policy, a critical aim is that of understanding the behaviour of rational utility-maximizing agents (for example, farmers), and how stimuli that reflect the costs and benefits of water may impact on their decisions. The immediate effect of water markets is that they lead to the realization of a transferable value of WALs by users. The first unintended consequence is that licences that were hitherto not used ('sleeper' licences) are suddenly found to be valuable, and sleeper WAL holders can profit from this endowment. Where monitoring is weak, water theft can further compound over-allocation problems (Gómez et al., 2017). The second unintended consequence is that since the reallocation of WALs 'typically involves the reallocation of return flows along with consumptive uses, trading may impair discharges and environmental and market uses downstream in the selling area' (Pérez-Blanco et al., 2020). This may happen due to higher consumption ratios and/or due to the reallocation of water away from the selling basin/catchment (for example, through a water transfer) (Huffaker, 2008).

Even where these two unintended consequences are accounted for and properly managed, there is concern regarding the asymmetric impacts and potential territorial imbalances of allocatively efficient water markets (for example, disappearance of water-intense traditional agricultural areas such as upstream rice fields due to water reallocation to more productive downstream areas, and related forward and backward linkages) (Hertel and Liu, 2016). Institutions in Italy and elsewhere in Europe have thus far been incapable of controlling for these third-party impacts. In the Italian context, addressing these issues calls not only for major reform of nationwide regulations, but particularly for a profound revision of, and better coordination among, regional WAL regimes. Despite some steps forward through, for example, the Permanent Observatory, progress has been limited (Mysiak et al., 2014). As a result, the recipe of full-fledged market-led developments is likely to result in externalities (at least in the short to medium run, until the issues above are addressed), which impair the principles of allocative efficiency upon which markets develop, and other negative impacts on third parties. Uncertainty regarding third-party impacts fuels opposition to trading in Italy and has resulted in the ban of water markets following a referendum. Italy could benefit largely from examples of successful water market implementation elsewhere (for example, California, Australia) to address these issues, although the thorough reform needed may take some time (see below for an example of WAL regime reform). Also note that, even if best practices in water market development and implementation are successfully replicated, this does not guarantee that negative impacts on third parties can be prevented. For example, in a recent workshop organized by the author in Venice on the topic of efficiency improvements through water reallocations, policy-makers and representatives from civil society expressed concern at some of the realized outcomes of water market implementation elsewhere, particularly those related to over-allocation issues that had to be addressed through further public disbursement (for example, perceived to be the case of buyback programmes in California and Australia). Participants in the workshop claimed that such an outcome would have a significant and potentially unbearable political cost in a country that rejected this formula explicitly with a landslide vote in a referendum.

A necessary condition (Wheeler et al., 2017) towards the development of water markets in Italy is the revision of its substandard regional WAL regimes (Santato et al., 2016). While recent efforts have helped to unravel some relevant water supply uncertainties in the face of climate change, relevant 'institutional arrangements necessary to enable efficient and equitable water trading' and 'to prevent over-allocation' (Wheeler et al., 2017) are missing in Italy. Young (2014) lists six WAL regime design principles for the successful implementation of water markets, namely: (1) unbundle water rights into their various components; (2) use one instrument for one goal, and never use the same instrument to pursue another objective; (3) define all entitlements and allocations in a manner that is consistent with the way that water is stored, flows across and through land; (4) minimize transaction costs over time to facilitate trade; (5) fully assign all the risks associated with a water use right to one interest group; and (6) robustness is needed, such as the definition of entitlements as shares that ensure that the only way one person's entitlement share can be increased is either to convince someone else to reduce their share-holding or to introduce a new resource into the system so that no one is worse off as a result of a change in the number of shares held. In two recent reviews of the Italian WAL regimes, Rey et al. (2018) and Santato et al. (2016) note that the PRBD and Italy overall comply with none of these principles.

In summary, transitioning towards a WAL regime compatible with trading in Italy requires: (1) a complete and centralized (basin-scale) WAL register; (2) metering; (3) predefined water (re)allocation rules during droughts that replace, or at least complement, the discretionary rules used in the Permanent Observatory; (4) defining entitlements as shares of usable resources; (5) compliance with hydrological integrity (for example, avoid appropriation of return flows) and integrated water resources management principles, notably through the effective implementation of river basin-scale management of water resources; (6) provided all the above-mentioned requirements are met, extending the duration of water entitlements to facilitate adaptation and efficient water use and (re)allocation; and finally, (7) strong public opposition against water markets in Italy and the fragmented competences in the Italian WAL regime advises the use of (temporary) additional clauses that limit trade so to prevent/reduce asymmetric impacts (Pérez-Blanco et al., 2017; Rey et al., 2018; Santato et al., 2016).

REFERENCES

AdBPo (2010). *Sintesi dell'analisi economica sull'utilizzo idrico. Elaborato 6, Piano di Gestione. Autorità di Bacino del Fiume Po.* Report 6, Autorità di Bacino del Fiume Po.
AdBPo (2016a). *Istituzione dell'Osservatorio Permanente sugli utilizzi idrici nel Distretto Idrografico del Fiume Po.* Memorandum of Understanding, Rome, Italy, Autorità di Bacino del Fiume Po. https://pianobilancioidrico.adbpo.it/wp-content/uploads/2017/06/Protocollo_Osservatorio_definitivo.pdf.
AdBPo (2016b). *Piano di Gestione del distretto idrografico del fiume Po.* River Basin Management Plan, Parma, Italy, Autorità di Bacino del Fiume Po. https://pianoacque.adbpo.it/piano-di-gestione-2015/.
Angelini, P., and Reiterer, M. (2014). *Guidelines for Climate Change Adaptation at the Local Level in the Alps.* Report, Milano, Italy, Italian Presidency 2013–2014 of the Alpine Convention. http://www.alpconv.org/en/publications/alpine/Documents/guidelines_for_climate_change_EN.pdf?AspxAutoDetectCookieSupport=1.
Brunetti, M., Buffoni, L., Maugeri, M., and Nanni, T. (2000). Precipitation intensity trends in northern Italy. *International Journal of Climatology*, *20*(9), 1017–1031.
Coppola, E., Verdecchia, M., Giorgi, F., Colaiuda, V., Tomassetti, B., and Lombardi, A. (2014). Changing hydrological conditions in the Po basin under global warming. *Science of the Total Environment*, *493*, 1183–1196.
EEA (2018). *Environmental Indicator Report 2017.* Copenhagen, Denmark, European Environment Agency. https://www.eea.europa.eu/airs/2017.
EEA (2019). *Environmental Indicator Report 2018.* Copenhagen, Denmark, European Environment Agency. https://www.eea.europa.eu/airs/2018.
Forzieri, G., Feyen, L., Rojas, R., Flörke, M., Wimmer, F., and Bianchi, A. (2014). Ensemble projections of future streamflow droughts in Europe. *Hydrology and Earth System Sciences*, *18*(1), 85–108.
Gómez, C.M., Pérez-Blanco, C.D., Adamson, D., and Loch, A. (2017). Managing water scarcity at a river basin scale with economic instruments. *Water Economics and Policy*, *4*(1), 1750004.
Groppelli, B., Soncini, A., Bocchiola, D., and Rosso, R. (2011). Evaluation of future hydrological cycle under climate change scenarios in a mesoscale Alpine watershed of Italy. *Natural Hazards and Earth System Sciences*, *11*(6), 1769–1785.
GU (2006). Decreto legislativo 3 aprile 2006, n. 152 Norme in materia ambientale.
GU (2008a). Decreto-legge 25 giugno 2008, n. 112.
GU (2008b). Legge 6 agosto 2008, n. 133.
GU (2009a). Decreto-legge 25 settembre 2009, n. 135, Disposizioni urgenti per l'attuazione di obblighi comunitari e per l'esecuzione di sentenze della corte di giustizia della Comunità europea.
GU (2009b). Legge 20 novembre 2009, n. 166.
GU (2009c). Legge 23 luglio 2009, n. 99, Disposizioni per lo sviluppo e l'internazionalizzazione delle imprese, nonché in materia di energia.
GU (2010). Sentenza n. 325 del 2010 della Corte costituzionale.
GU (2011). Abrogazione parziale, a seguito di referendum popolare, del comma 1 dell'articolo 154 del decreto legislativo n. 152 del 2006, in materia di determinazi-

one della tariffa del servizio idrico integrato in base all'adeguata remunerazione del capitale investito.

Hertel, T.W., and Liu, J. (2016). *Implications of Water Scarcity for Economic Growth.* OECD Environment Working Papers, Paris, Organisation for Economic Co-operation and Development. http://www.oecd-ilibrary.org/content/workingpaper/5jlssl611r32 -en.

Huffaker, R. (2008). Conservation potential of agricultural water conservation subsidies. *Water Resources Research, 44*(7), W00E01.

IPCC (2018). Impacts of 1.5°C of global warming on natural and human systems. In *Global Warming of 1.5°C. An IPCC Special Report on the Impacts of Global Warming of 1.5°C above Pre-Industrial Levels and Related Global Greenhouse Gas Emission Pathways, in the Context of Strengthening the Global Response to the Threat of Climate Change, Sustainable Development, and Efforts to Eradicate Poverty.* Geneva, Switzerland, Intergovernmental Panel on Climate Change (IPCC). https://www.ipcc.ch/site/assets/uploads/sites/2/2018/11/SR15_Chapter3_Low_Res .pdf.

Italian Ministry of Environment, Land and Sea (2013). *Sixth National Communication under the UN Framework Convention on Climate Change.* Communication, Rome, Italy, Ministry of Environment, Land and Sea. https://unfccc.int/files/national _reports/annex_i_natcom/submitted_natcom/application/pdf/ita_nc6_resubmission .pdf.

MATTM (2014). *Strategia Nazionale Di Adattamento Ai Cambiamenti Climatici Indice (National Strategy of Adaptation to Climate Change).*

MATTM (2017). *Piano Nazionale Di Adattamento Ai Cambiamenti Climatici PNACC. Prima Stesura per La Consultazione Pubblica Luglio 2017.*

Montanari, A. (2012). Hydrology of the Po River: looking for changing patterns in river discharge. *Hydrology and Earth System Sciences, 16*(10), 3739–3747.

Mysiak, J., Carrera, L., Amadio, M., Pérez Blanco, C.D., and Santato, S. (2014). *Development of MSPs in the Po River Basin District. Controlled Floods on Agricultural and Scarcely Developed Rural Land (MSP-F) and Managing Severe Drought Spells in the Otherwise Water-Abundant River Basin District (MSP-D).* Deliverable 7.2, Fondazione Eni Enrico Mattei.

OECD (2013). *Water and Climate Change Adaptation: Policies to Navigate Uncharted Waters.* Report, Paris, France, Organisation for Economic Co-operation and Development, www.oecd.org/env/resources/waterandclimatechange.htm.

OJ (2000), Water Framework Directive 2000/60/EC.

Pérez-Blanco, C.D., Essenfelder, A.H., and Gutiérrez-Martín, C. (2020). A tale of two rivers: integrated hydro-economic modeling for the evaluation of trading opportunities and return flow externalities in inter-basin agricultural water markets. *Journal of Hydrology, 584*, 124676.

Pérez-Blanco, C.D., Koks, E.E., Calliari, E., and Mysiak, J. (2017). Economic impacts of irrigation-constrained agriculture in the Lower Po Basin. *Water Economics and Policy, 4*(1), 1750003.

Pineschi, G., and Colaizzi, M. (2014). La definizione dei costi ambientali e della risorsa. Presentation presented at the Workshop Governance e analisi economica, Bologna.

Ravazzani, G., Ghilardi, M., Mendlik, T., Gobiet, A., Corbari, C., and Mancini, M. (2014). Investigation of climate change impact on water resources for an alpine basin in Northern Italy: implications for evapotranspiration modeling complexity. *PLoS ONE, 9*(10), e109053.

Regione Piemonte (2017). *Condizionalità ex ante riferita al settore delle risorse idriche prevista dall'Accordo di Partenariato ai fini dell'accesso ai fondi europei relativi al Programma di Sviluppo Rurale 2014–2020.* Attuazione della DGR n. 43-4410 del 19 dicembre 2016.

Rey, D., Pérez-Blanco, C.D., Escriva-Bou, A., Girard, C., and Veldkamp, T.I.E. (2018). Role of economic instruments in water allocation reform: lessons from Europe. *International Journal of Water Resources Development, 35*, 206–239. https://doi.org/10.1080/07900627.2017.1422702.

Santato, S., Mysiak, J., and Pérez-Blanco, C.D. (2016). The water abstraction license regime in Italy: a case for reform? *Water, 8*(3), 1–15.

Servizio Studi – Dipartimento ambiente della Camera dei Deputati (2011). *Il Servizio Idrico Integrato e il Referendum Abrogativo del 12–13 Giugno 2011.* Report, Rome, Italy, Camera dei Deputati. http://documenti.camera.it/leg16/dossier/Testi/AM0232.htm.

Tibaldi, S., Cacciamani, C., and Pecora, S. (2010). Il Po nel clima che cambia. *Biologia Ambientale, 24*, 21–8.

Vezzoli, R., Mercogliano, P., Pecora, S., Zollo, A.L., and Cacciamani, C. (2015). Hydrological simulation of Po River (North Italy) discharge under climate change scenarios using the RCM COSMO-CLM. *Science of the Total Environment, 521–522*, 346–358.

Wheeler, S.A., Loch, A., Crase, L., Young, M., and Grafton, R.Q. (2017). Developing a water market readiness assessment framework. *Journal of Hydrology, 552*(Supplement C), 807–820.

Young, M.D. (2014). Designing water abstraction regimes for an ever-changing and ever-varying future. *Agricultural Water Management, 145*, 32–38.

Zanchettin, D., Traverso, P., and Tomasino, M. (2008). Po River discharges: a preliminary analysis of a 200-year time series. *Climatic Change, 89*(3–4), 411–433.

11. Applying the WRMA framework in England

Rosalind H. Bark and Nancy E. Smith

11.1 INTRODUCTION

11.1.1 Water Markets in the UK

Each nation in the United Kingdom (UK) manages water differently and here we apply the water market readiness assessment (WMRA) framework (Wheeler et al., 2017) to England. In England, the key regulatory agency responsible for managing water resources is the Environment Agency (EA) and the principal mechanisms used are surface water and groundwater abstraction licences and rules around their allocation, use and trade. In this context, water trading is the trade of abstraction licences (EA, 2011a).

Distinctive from many other countries that trade water rights, where the predominant use is irrigated agriculture (Wheeler et al., 2015), in England the predominant use is for public water supply (PWS). There are nine PWS companies operating as regulated monopolies providing drinking water and wastewater services. A further 12 water companies supply drinking water only. PWS companies are regulated by five agencies: the EA, which enforces abstraction licences, environmental standards and return flow quality; Ofwat, the economic regulator; the Drinking Water Inspectorate (DWI), the water quality regulator; the Consumer Council for Water, which represents the interest of water consumers; and the UK Government Department for Environment, Farming and Rural Affairs (Defra), which is responsible for setting policy and legislation to maintain and improve water quality to meet, at the time of writing, European Union (EU) Directives, for example the Water Framework Directive (EU WFD).

Other water users must also apply for abstraction licences if their collective water use is over 20 m³/day. While there are regional differences, the second-largest abstractor is often the energy sector, followed by other industrial users. Agriculture represents on average 1 per cent of abstraction across England and Wales (Lumbroso et al., 2014). Nevertheless, increasing water stress with climate change, population growth and the seasonal demands of the energy, agricultural and PWS sectors, combined with the needs of the environment, provide the impetus to further develop water trading.

The predominance of the PWS sector makes it central in any water trading discussion. This chapter was written at the end of the PWSsector's 2015–19 Asset Management Period (AMP6) and Price Review (PR19), and at the commencement of AMP7 and PR21. To coincide with the AMP, PWS companies submit water resource management plans (WRMPs). A review of the 2019 WRMPs reveals that water trading is one of the preferred supply management options. However, this is primarily in the context of intra and inter-PWS company water transfers and not inter-sectoral water trade. Such transfers are supported by Ofwat; see Box 11.1.

BOX 11.1 OFWAT AND WATER TRADING

Water trading is where a water company responsible for supplying water in an area buys it from another water company rather than developing its own water resources. Since privatisation, the level of trading has remained static at around 4 to 5 per cent of water into supply, however, at the same time, water companies have invested heavily in linking up their own networks and moving water internally. Greater levels of water trading can benefit customers through improvement to resilience of supply, and the environment, by ensuring water is moved from areas of surplus to areas where water is scarce. In our impact asessment to support our Water 2020 reforms we identified that greater water trading between water companies could save up to £600 million over 30 years (2015/16 prices). (Ofwat, 2018)

The slow uptake of inter-sector water trading is in part a response to the regulatory environment (EA and Ofwat, 2009) and the requirement for PWS companies to make a business case to Ofwat to invest in innovative activities (Tinch, 2009).

Against this backdrop, periods of water stress are becoming more frequent and it is likely that there will be greater willingness to explore water markets. This is reflected in academic research sponsored by Defra, UK government funders, PWS companies, and the UK Water Industry Research (UK WIR). This research has explored the relevance of water trading in the UK with lessons from Australia (Defra, 2013), modelled various trading options (Erfani et al., 2014, 2015) and explored stakeholders' attitudes and readiness to participate in a water market (Lumbroso et al., 2014). Nevertheless, following this period of active research the practical uptake of options was limited during AMP6.

11.1.2 Water Reforms and Trading Alternatives in the UK

In 1989, the PWS sector in England was privatised. Yet, the regulatory environment constrains the ability of privatised PWS companies to use market-based instruments (MBIs). Some are gaining experience with MBIs; for example, Anglian Water (which operates in Eastern England) developed a set of catchment level incentives to reduce a pollutant at source to meet EU WFD provisions (Mohamad Ibrahim et al., 2019). It is also using innovation, through direct engagement with its customers, to drive sustainable change including demand management (Anglian Water and Davies, 2018; Hoolohan and Browne, 2018).

Alternatives and enhancements to water trading are also being explored by PWS companies, for example through Regional Water Resource Management Groups (RWRMGs). The immediate impetus for the acceleration of regional water resource planning (RWRP) is the 2018 joint letter from Defra to the PWS sector (Defra, 2018). The joint letter set five challenges, of which three focus on PWS activities, specifically through a call for more ambition from PWS in the development of RWRMGs; to build resilience through RWRP; and to adopt greater use of markets and competition. A key objective of the RWRMGs is to enhance water resource management through collaborative working across all water use sectors. There are five RWRMGs at various stages of development; see Figure 11.1. Water Resources East (WRE) was the first RWRMG to gain independence from its PWS company host. WRE covers over 31 000 km² in Eastern England (see Figure 11.2).

WRE is supportive of water trading (see Box 11.2). Its regional planning efforts are underpinned by a technical programme consisting of: sub-regional planning, abstraction management reform, environmental land management schemes, and incentive-based pilots working directly with landowners.

BOX 11.2 WRE VIEW ON WATER TRADING

WRE advocates for the advancement of multi-sector water sharing in Eastern England as part of its RWRP for a more flexible and adaptive use of available water; this includes water trading (WRE, 2020).

To support future water trading, researchers at the University of Manchester and WRE are developing a simulator-based trading tool. This builds on existing water-sharing and trading research on catchments in

Eastern England during WRE's phase one regional strategy (2014–18) and scoping research on the Bedford Ouse catchment (University of Cambridge and Anglian Water, 2013). WRE is also working in partnership to develop a proof of concept for an agricultural water trading platform Wheatley Watersource which can accelerate both water rights and post-abstraction trading in times of need.

Source: WRE (2020), map design by Avidd Design.

Figure 11.1   RWRMG in England

Figure 11.2   WRE geographical area

11.2   WMRA FRAMEWORK

In this section we apply the WMRA framework to a single catchment in Eastern England, the Upper Ouse and Bedford Ouse. A series of research articles have focused on this catchment (Erfani et al., 2014, 2015; Lumbroso et al., 2014). The catchment has around 205 abstraction licences (Lumbroso et al., 2014), of which 94 are active (Erfani et al., 2014).

It provides a good test catchment for a number of reasons: it is located in one of the driest regions of England with abstraction reliability less than 30 per cent (EA, 2017); the catchment itself is drier in the east and wetter in the west (Erfani et al., 2014); and there are four large water user groups: a PWS company – Anglian Water – with multiple reservoir storage units, an energy supplier with cooling requirements, local industry, and a large agricultural sector. Furthermore, there are several Sites of Special Scientific Interest (SSSIs), Ramsar wetlands, and Areas of Outstanding Natural Beauty (AONBs) with freshwater-dependent ecosystems. These characteristics present the fundamental elements for water trading viability: water stress, spatially distributed water users that have different seasonal water use, and different marginal value of water between sectors.

11.2.1   Stage One: Necessary Conditions: Water Access and Allocation Arrangements

Hydrology

Documented hydrology system
We cannot address the comprehensiveness of hydrologic data for every catchment in England. However, there is a lot of publicly available data on UK-wide hydrological systems as recorded by the British Hydrological Society (2014), such as data from the British Geological Survey, UK Met Office, Centre for Ecology and Hydrology, and the EA.

Understanding of connected systems
Within the study catchment, and more generally, there is good understanding of connected systems. This data supports EA abstraction licence trading rules that permit trading within a surface water catchment, groundwater aquifer and, in some instances, between a linked surface and groundwater system (EA, 2011a). However, in order to develop trading system knowledge, it is likely in the first instance that any new trading would be intra-catchment. This is because inter-catchment trading carries potentially more externality risks and

greater modelling and regulatory capability requirements than is currently possible.

Future impacts modelled

Defra has undertaken a series of studies to better understand unmet demand and the viability of trading, particularly in catchments where there is water stress and seasonal or permanent unmet demand (Defra, 2011a). On longer timescales, Defra has modelled the impacts of climate change and population growth on supply and demand, and these projections have accelerated interest in developing water trading (Defra, 2011b). At the PWS company level, each PWS company must develop asset management strategies to meet water resource deficits. At the RWRMG level, RWRP extends beyond PWS to model asset management strategies to meet future water needs for all water users in a region (e.g., Moncaster, 2018 in WRE, 2020).

Trade impacts understood

The EA must approve all abstraction licence trades. Three key restrictions for trades are the consideration of hands-off (environmental) flows (HoFs), land conditions, and rules defined under Section 57 of the Water Resources Act, 1991. HoFs are conditions on abstraction licences to ensure minimum flows in rivers; land conditions stipulate the area or purpose that water is to be used on; and Section 57 is a drought emergency provision to restrict irrigation. In England, one-quarter of abstraction licences are subject to HoF restrictions, representing 40 per cent of surface water licences and 3 per cent of groundwater licences. HoF restrictions do not apply to 'legacy licences'; granted pre-privatisation and between the 1963 and 2003 Water Acts (Cunningham, 2002; EA and Ofwat, 2012; Lumbroso et al., 2014), but this will be reviewed under the announced abstraction reform process.

Resource constraints understood

The EA classes 18 per cent of river catchments in England as over-licensed, 15 per cent as over-abstracted (EA, 2008), and it reports that a quarter of new abstraction licences do not provide reliable water supplies (EA, 2011b). At the PWS company level, as part of the AMP process, demand and supply imbalances must be identified. RWRMGs are also building capacity to understand resource constraints. For instance, WRE has identified a combination of risks including forecasting population growth and climate change impacts on previously predominately rain-fed agriculture, the energy sector and vulnerable ecosystems (WRE, 2021).

Resource constraints enforced
Water resource constraints in catchments are enforced through abstraction caps. During drought conditions restrictions on use apply first to the newest licences that have been issued. Older, unrestricted licence holders can continue to abstract until formal drought orders are imposed. This can lead to unintended conflicts amongst abstractors in times of water stress. In response, in 2017 the EA introduced an 'essential water need' as part of ongoing abstraction reform. This allows the EA to override pre-existing licence rules and enforce restrictions on all abstractors. Exemptions to meet this essential water need only apply if the restrictions have public health and/or social impacts, such as to PWS and electricity supply disruption (Food and Drink Federation, 2017).

In our study catchment, the EA has developed and implemented an abstraction licensing strategy to manage *in situ* groundwater and downstream impacts from abstractions (EA, 2017). As part of the licensing strategy, no new licences will be issued, and licence renewals are based on considerations of environmental sustainability, justification of need and water efficiency. Groundwater licences in the catchment are fully used and therefore new entrants must either buy or lease land with a licence (Erfani et al., 2014).

In extreme circumstances, the EA has the option to prohibit all future abstraction permanently. For instance, from 2021 the groundwater abstraction licences of the PWS company (Anglian Water) and local farmers in Ludham, Norfolk will be revoked in order to protect the Norfolk Broads fenland habitat from further desiccation. This situation has created years of conflict between water users and the EA, and has demonstrated the need for greater coordination between water users. Anglian Water will construct a pipeline to Ludham to meet PWS needs, while farmers and landowners must look to diversify crops or implement water efficiency measures (EA, 2018; Hill, 2019a, 2019b; RSPB, 2019).

Water use enforced
We saw above that to limit externality impacts on vulnerable ecosystems the EA has implemented the abstraction licensing strategy in our study catchment, and will revoke groundwater licences in Ludham, Norfolk. This case has highlighted concerns over EA processes, with the National Farmers' Union (NFU) believing that local farmers were not provided adequate notice to find alternative water sources or to otherwise adapt to the change (Hill, 2019a).

System type

Suitability of water sources for trade
We have seen that the EA abstraction licencing rules do allow for trading of surface water and groundwater within linked systems, but only if certain

rules and requirements are met. In practice, however, more modelling is likely needed, especially around groundwater flows, its interaction with surface water abstraction and issues around conjunctive use, to ensure the viability of surface water and groundwater trading (Erfani et al., 2014, p. 4742).

Transfer infrastructure availability/suitability
As a mostly privatised industry there are several owners of key water transfer and supply infrastructure, *inter alia*, farmers and landowners (private reservoirs, farm-to-farm pipe transfers), PWS companies (water treatment and recycling plants, reservoirs, pipelines and delivery infrastructure), and environmental non-governmental organisations (such as river and wildlife Trusts) and local authority bodies (canal systems, reservoirs, wetland areas, nature reserves). Floodwater pumping infrastructure is owned by the Internal Drainage Boards that manage them. In this system of public–private ownership, it can make the possibility of water trading more complex.

Through the AMP process, some PWS companies, for example Anglian Water, have committed to constructing a network of strategic pipelines to transfer surplus water resources, primarily intra-PWS between their own Water Resource Zones and inter-PWS (Anglian Water, 2019) but also potentially inter-sector through existing waterways to increase connectivity and resilience (WRE, 2021). Intra-PWS transfer infrastructure is straightforward and encouraged by the regulator (see Box 11.1).

Regulation requirements for trade
In the period 2003–08, other than the short-term water trading facilitated by the EA during the 2018 agricultural drought, only 48 official trades were completed in England and Wales, of which 31 were in the East of England (Erfani et al., 2014). The entire licence, or part of the licence, can be traded on a permanent or temporary basis. The fee for approved trades is only £135 but it can take three or four months to be approved (EA, 2008, 2011a). Consideration of a temporary licence takes around 28 days. Most approved trades are 'pair-wise' between a willing seller and buyer. Details of the trade, for example the price, are not shared but recent quantitative research has revealed low prices, with many trades described as 'water sharing' instead of formal 'trading' (Tabas et al., 2019).

Property rights and institutions

Water legislation
Recent papers discuss key water legislation (Erfani et al., 2014, 2015; Lumbroso et al., 2014). Additionally, Lumbroso et al. (2014) provide an outline of abstraction licensing systems in England. In brief, the EA determines

the maximum abstraction in each catchment, and any abstractors diverting, collectively, more than 20 m³ of water daily must have an abstraction licence (EA and Ofwat, 2012). The current system favours 'grandfathered' abstraction licences,, and licences granted in perpetuity. In the 2010s, around 80 per cent of the 21 000 abstraction licences in England and Wales (Lumbroso et al., 2014) were 'grandfathered' licences (Frontier Economics and Anglian Water, 2011). Furthermore, in England around a third of abstraction licences are unused, for example 'sleeper' licences, representing 20 per cent of the licensed volume (EA and Ofwat, 2012). The EA is reviewing all grandfathered and sleeper licences and is considering buying them back or reducing the allocation based on actual use (Tabas et al., 2019).

Unbundled rights
As per the Water Act 1963, permanent licences were based on riparian rights. These licences were based on three years of self-reported water abstraction volumes and granted in perpetuity, without a determination of any externalities imposed on other users or the environment (Lumbroso et al., 2014). A consequence is that potential new abstractors have difficulty entering the market in over-licensed or over-abstracted catchments, and are often restricted to buying or leasing land that has an existing abstraction licence (Lumbroso et al., 2014; Tabas et al., 2019). New licences, to align with the abstraction reform process, are time-restricted and granted for between six and 18 years. At renewal, an additional 12 years is normally granted (EA, 2014).

Rights transferable
Abstraction licences are transferable on a temporary or permanent basis. Although the EA's administrative fees are only £135, the approval process that considers potential environmental externalities and impacts on other users from trade can take many months and may involve the need to employ experts (CIWEM, 2014; Lumbroso et al., 2014). Informal pre- and post-abstraction trades within the agricultural sector illustrate irrigators' use of workarounds (Tabas et al., 2019) to circumvent the long timescales involved in trading abstraction licences. These workarounds may also reveal a misalignment between the trading system and the short-term demand for trades (NFU and Cranfield University, 2018).

Constraints between connected systems
In volume terms, the largest water transfers are between PWS companies. Most of these bulk water transfers were agreed pre-privatisation and are in perpetuity. For example, in 2017–18, Anglian Water received 18 transfers of a total 1 084 006 m³ of treated water from five other PWS companies (Ofwat, 2019). Ofwat is keen to expand inter-regional trade between PWS companies;

however, the EA remains cautious because of the invasive species risk, as well as potential risks to PWS water security and the environment (CIWEM, 2014).

11.2.2 Stage Two: Developing Water Markets

Market evaluation, development and implementation
A key aspect to developing water markets is robust modelling of potential trades and trading rules. Erfani et al. (2014) investigate pair-wise spot market trading using an optimisation-based hydro-economic model that considers environmental constraints, consumptiveness of use, transaction costs and the estimated marginal value of water for different users. In their modelling, they find that the energy and agricultural sectors, which face abstraction restrictions in times of stress (that is, hands-off flow rules), would benefit most from short-term trading. The same authors also model different environmental flow regimes (Erfani et al., 2015). In the current system, abstraction licences are subject to minimum environmental flow restrictions which, when reached, lead the EA to suspend abstraction licences. The modelling suggests that a more dynamic and equitable environmental flow licencing system would more effectively protect river flows during drought, but with potential dry-year losses in economic benefits of 10–15 per cent.

Another aspect of water trading development is stakeholder acceptance. In water market demonstration workshops, Lumbroso et al. (2014) trialled two different trading systems: an improved pair-wise trading system, and a multi-lateral common pool market with a formal bidding process and a third-party catchment manager. Stakeholders engaged in the workshops but did not yet view water trading as an urgent issue.

In 2018, the summer drought (conditions were the driest since 1921; CEH, 2018) triggered limited implementation of water trading. The EA facilitated temporary water trades in the Eastern England agricultural sector. It launched a web-based mapping platform identifying areas where potential water trades would be permitted. This online platform complemented the EA's 'flexible licensing' position during the drought. Concurrently, the NFU created a web-based 'Water Bank' to match buyers and sellers (NFU, 2018). The EA platform has been extended to three more regions in England, and the NFU remains committed to assisting its members with trading.

Externalities, governance, adjustment, entitlement/rights registers and system type

Strong governance impartiality
Lumbroso et al. (2014) note that once the EA has given approval for a trade it is impartial during that trade. However, the perception of impartiality is not

universal. A reason raised for the small number of water trades, other than the costs of experts, the slow approval process and the difficulty in finding a trading partner, is how the EA will treat trade in sleeper licences. Farmers who hold a large number of sleeper licences expressed reservations of involvement in short- or long-term trade, out of fear that the EA will take the opportunity of the trade to restrict such previously inactive licences.

Existence of externalities understood
Erfani et al. (2015) summarise information regarding environmental and water security externalities. On the ground, the EA plan to revoke groundwater abstraction licences in Ludham, Norfolk is a response to the externality impacts of current water use on the Norfolk Broads fenland habitat (EA, 2018). There is also concern that the large number of sleeper licences represents a potential externality trade threat (Erfani et al., 2015).

Water use monitored
As per the Water Act of 1963, 'Most abstractors pay operational licence fees based on the size of their licence, not on how much water they actually use' (Lumbroso et al., 2014, p. 2679). This lack of water use monitoring is a serious constraint to water market development. However, the EA does have information on the 'consumptiveness' of different water use sectors (referenced in Lumbroso et al., 2014, to determine the ease of pre-approved trades).

11.2.3 Stage Three: Monitoring and Continuous Review/Assessment

Adjustment

Gains from trade (no. users/transaction costs/diversity of use)
It was previously noted that in modelling pair-wise trading, Erfani et al. (2014) find that the energy and agricultural sectors which face abstraction restrictions in times of stress (HoF rules) benefit most from short-term trading. Of the trading simulated, more than 90 per cent of the trades by volume in normal and dry years were from the PWS companies to the energy sector. An explanation is that the buyers – that is, the energy sector – favour sellers who can supply large volumes to minimise transaction costs. Despite low marginal values in the agricultural sector, more than 90 per cent of trades were intra-sector. The authors also modelled the effect of higher fixed transaction costs on trading: the number of trades diminishes in both normal and dry years, particularly between agricultural abstractors. Finally, they also modelled feedbacks between transaction costs, trading volumes and water abstraction from the river and a PWS reservoir. They conclude that the short-term water market increases net benefits across all water use sectors.

In Erfani et al. (2015) they extend their modelling to compare the current system based on volumetric abstraction licences with fixed minimum flow requirements (HoF rules) with a share-based licensing system with dynamic environmental minimum flows. Modelling the current system, they report large variability in annual river flows, including zero flows, that can be attributed to grandfathered abstraction licences that have no HoF restrictions. In contrast, their alternative system results in less variable river flows and no zero flows, and more active use of reservoir storage. Securing environmental flows has an opportunity cost, with estimated economic losses of 15 per cent compared to the status quo with a water market.

In this alternative system, the volume of water traded almost doubles and the number of trades more than doubles. The end of HoF rules (which do not apply to the PWS licences) results in reduced trading from the PWS to the energy sector, and the agricultural sector plays a more important role selling water to the energy sector from April in any given year and to the PWS company after the growing season. Later, in the autumn, the energy sector sells to the PWS company. In a dry year, over the 12 months each sector both buys and sells water.

Political acceptability of trade

Water resources management is challenged by climate change, environmental concerns, increasing demand and potential new entrants (Erfani et al., 2015) which, when combined, have partly driven regulator interest in policy reforms, including around water trading (Lumbroso et al., 2014). Modelling through predicted trading behaviour is critical to the acceptability of further water trade development (Erfani et al., 2015).

In 2018, all regulators sent a joint letter to all PWS companies to demonstrate their support for PWS activities in the areas of water transfers, regional water planning and the use of water markets (Defra, 2018). In turn, the regulators committed to develop a more responsive regulatory approach. In 2018/19, the Draft National Policy Statement for Water Resources Infrastructure was open to consultation (Defra, 2019) and in September 2020 Defra released a 'Water abstraction plan' policy paper and timetable for abstraction reform (Defra, 2020). The EA is carrying out abstraction licence reform and is leading the implementation of a National Framework for water resources.

Another aspect of political acceptability is the appetite among catchment stakeholders for trade. As introduced above, Lumbroso et al. (2014) investigated stakeholder attitudes to trade using a mixed-method approach including two interactive demonstration workshops. Stakeholders were introduced to, and then 'traded' on, a weekly time-step in two markets: an improved pair-wise system and a common pool trading market. In general, the stakeholders were willing to consider water trading, though few perceived it as an 'urgent' issue

(Lumbroso et al., 2014). They were most comfortable with improved pair-wise trading, the option closest to the status quo. Stakeholders approved of two market characteristics: transparency that revealed 'market' prices; and flexibility to undertake fast-tracked (pre-approved, and therefore less expensive) short-term trading, which should lower transaction costs and perceived needs to game the system with workaround solutions.

Stakeholders raised concerns about any changes to the reliability of their abstraction licences, around fairness of curtailment rules under dry conditions that might favour those users with long-term water security needs such as PWS companies and environmental uses, and the potential negative impact of water trade on rural communities. In response, they wished to 'ring-fence' sectors in order to limit the power asymmetry of big players in any new system. Other feedback related to the indirect role of trading in the development of relationships between abstractors, and in fostering collective responsibility for water resources management. There was an expressed desire to be engaged in any future market development.

Entitlement registers and accounting systems

Trustworthy systems
Lumbroso et al. (2014), in their assessment of stakeholders' attitudes towards trading, uncover a need for learning, not only on the side of potential buyers and sellers, but also on the side of the regulators. The regulators need to build confidence in the models that underpin any trading scheme and in market rules around fairness and the protection of smaller players, rural communities and the environment.

Trade and market information availability
From their interaction with stakeholders Lumbroso et al. (2014, p. 2692) conclude that there is a 'widespread lack of knowledge and awareness of water trading across all types of stakeholders in England'. Indeed, most did not know the value of water, but perceived it to be low. Water trade facilitated in 2018 by the EA and NFU (NFU, 2018), as well as the year-long trial and expansion of a new collaborative water trading platform in Eastern England (Wheatley, 2021), will in time generate more trade and market information.

11.3   CRITICAL LESSONS FROM THE UK

11.3.1   WMRA Stages or Enabling Conditions?

We suggest that the WMRA steps are not necessarily sequential; that is to say, it may not be necessary to have completed one stage to progress to the next stage. There are many enabling conditions for water markets to develop, and these may develop in parallel. Indeed, we see that there is a portfolio of efforts required to strengthen water market readiness, such as wider abstraction licence reform, investment in transfer infrastructure, and research to model potential trades or trading systems and the acceptability of trading options to stakeholders. We suggest also that there are pivotal roles for other agents, for instance environmental non-governmental organisations and PWS companies, to promote and normalise discussions around water markets. More generally, we saw that market-based instruments could be viewed as a precursor to creating conditions for water users and regulators alike to consider new approaches.

11.3.2   Water Markets Might be Informal

Stage 2 of the WMRA concerns the development of water markets, but in our study catchment there is evidence of a dynamic and functional informal market based on land trades. This practice is a pre-abstraction water trade, as the abstraction licence is tied to the land (Tabas et al., 2019). Additionally, there were instances of informal post-abstraction water transfers between farmers using reservoir infrastructure. Lessons learned from this informal market can inform and strengthen the future development of a formal water market.

11.3.3   Direction of Travel

The direction of travel is strong support from the regulators and the PWS companies for abstraction licence reform and water market development. However, most of the research is focused on a single catchment and it is unlikely to be a blueprint for water markets more generally. As other markets develop, regulators will need the capacity to respond to more localised and regional trading systems.

11.3.4   Models and Learning

A noted barrier to short-term spot trading is the time the EA takes to assess environmental externalities and third-party impacts from trade. The informal market in the agricultural sector subverts this approval process. Nevertheless, this focus on time, we believe, overlooks a critical need for trusted decision support tools. Key lessons from countries with active water markets, for example the United States and Australia, suggest that tools work best when stakeholders perceive them as accurate, transparent and representative of diverse interests (Wheeler et al., 2018). This wider focus on decision-making tools and learning from the experience in other countries could be the first step in overcoming any cultural reluctance to water trade expressed in a desire to 'ring-fence' sectors (Lumbroso et al., 2014). Other steps could include a different framing for water trade; for instance, framing it as water sharing (Tabas et al., 2019) or emphasising that all sectors can be buyers and sellers (Erfani et al., 2015).

There are signs that a new round of research is forthcoming. We recommend that this new research agenda works to inform the next steps in water resource management reforms. RWRMGs could play a pivotal role in creating networks between sectors and sharing new learning and good practice from both domestic experience in other utilities – that is to say, energy and emissions trading – and overseas experience with water markets. A multi-sector, abstraction licence pilot market (pre- and post-abstraction) would be invaluable.

REFERENCES

Anglian Water (2019). *Water Resource Management Plan 2019*. www.anglianwater.co
.uk/siteassets/household/about-us/wrmp-report-2019.pdf.

Anglian Water and Davies, W. (2018). *New Models for Collaborative Working: A Guide to Innovation from Our Shop Window*. www.bitc.org.uk/sites/default/files/ anglian_water_shop_window_toolkit_aw_sw.pdf.

British Hydrological Society (2014). Data sources. www.hydrology.org.uk/Data _sources.php.

Centre for Ecology and Hydrology (CEH) (2018). UK hydrological status update – early August 2018. www.ceh.ac.uk/news-and-media/blogs/uk-hydrological-status -update-early-august-2018.

Chartered Institution of Water and Environmental Management (CIWEM) (2014). *Policy Position Statement: Bulk Water Transfers*. www.ciwem.org/assets/pdf/ Policy/Policy%20Position%20Statement/Bulk-Water-Transfers.pdf.

Cunningham, R. (2002). Reform of water resource control in England and Wales. *Journal of Water Law*, *13*(1), 35–44.

Defra (2011a). *Characterising Potential Water Abstraction Licence Markets: Phase 2 – Catchments for Trading Promotion*. Department for Environment, Food and Rural Affairs, London, UK.

Defra (2011b). *Characterising Potential Water Abstraction Licence Markets*. Department for Environment, Food and Rural Affairs, London, UK.

Defra (2013). *Detailed Case Study of the Costs and Benefits of Abstraction Reform in a Catchment in Australia with Relevant Conditions to England and Wales*. R&D Technical Report WT1504/TR, Department for Environment, Food and Rural Affairs, London, UK.

Defra (2018). Building resilient water supplies – a joint letter. www.ofwat.gov.uk/wp -content/uploads/2018/08/Building-resilient-water-supplies-letter.pdf.

Defra (2019). Consultation on the draft National Policy Statement for Water Resources Infrastructure. https://consult.defra.gov.uk/water/draft-national-policy-statement/ (accessed 24 March 2021).

Defra (2020). Policy paper: Water abstraction plan. https://www.gov.uk/government/ publications/water-abstraction-plan-2017/water-abstraction-plan#summary (accessed 24 March 2021).

Environment Agency (EA) (2008). *Water Resources in England and Wales – Current State and Future Pressures*. Environment Agency, Bristol, UK.

Environment Agency (EA) (2011a). *A Guide to Water Rights Trading*. Report – GEHO0711BTZK-E-E, Environment Agency, Bristol, UK.

Environment Agency (EA) (2011b). *The Case for Change – Current and Future Water Availability*. Environment Agency, Bristol, UK.

Environment Agency (EA) (2014). Abstracting water: a guide to getting your licence. Environment Agency, Bristol. https://assets.publishing.service.gov.uk/government/uploads/system/uploads/attachment_data/file/716363/abstracting-water-guide-to-getting-licence.pdf (accessed 24 March 2021).

Environment Agency (EA) (2017). *Upper Ouse and Bedford Ouse Abstraction Licensing Strategy: A Strategy to Manage Water Resources Sustainably*. Report – 227_10_SD01 version 7, Environment Agency, Bristol, UK.

Environment Agency (EA) (2018). *Catfield Fen: Decision on Licence Applications*. www.gov.uk/government/publications/catfield-fen-decision-on-licence-application/catfield-fen-decision-on-licence-applications#contents.

Environment Agency (EA) and Ofwat (2009). *Review of Barriers to Water Rights Trading*. Report –GEHO0109BPKR-E-E, Environment Agency, Bristol, UK.

Environment Agency (EA) and Ofwat (2012). *The Case for Change: Reforming Water Abstraction Management in England*. Report – GEHO1111BVEQ-E-E, Environment Agency, Bristol, UK.

Erfani, T., Binions, O., and Harou, J.J. (2014). Simulating water markets with transaction costs. *Water Resources Research*, 10.1002/2013WR014493.

Erfani, T., Binions, O., and Harou, J.J. (2015). Protecting environmental flows through enhanced water licensing and water markets. *Hydrology and Earth System Sciences*, *19*, 675–689.

Food and Drink Federation (2017). *Abstraction Reform: The Case for Prioritising Water for Food Production as an 'Essential Water Need'*. www.fdf.org.uk/keyissues-abstraction-reform.aspx.

Frontier Economics and Anglian Water (2011). A right to water? Meeting the challenges of a sustainable water solution. https://www.anglianwater.co.uk/siteassets/household/about-us/a-right-to-water.pdf (accessed 24 March 2021).

Hill, C. (2019a). Major new pipeline planned from Norwich to Ludham to alleviate Broads water fears. *Eastern Daily Press*. www.edp24.co.uk/news/environment/anglian-water-plans-new-pipeline-norwich-to-ludham-1-5985034.

Hill, C. (2019b). 'The Broads are being destroyed before our eyes', says Catfield Fen campaigner. *Eastern Daily Press*. www.edp24.co.uk/business/farming/catfield-fen-owner-joins-ea-abstraction-review-debate-1-5966225.

Hoolohan, C., and Browne, A.L. (2018). Reimagining spaces of innovation for water efficiency and demand management: an exploration of professional practices in the English water sector. *Water Alternatives*, *11*(3), 957–978.

Lumbroso, D.M., Twigger-Ross, C., Raffensperger, J., Harou, J.J., Silcock, M., and Thompson, A.J.K. (2014). Stakeholders' responses to the use of innovative trading systems in East Anglia, England. *Water Resources Management*, *28*, 2677–2694.

Mohamad Ibrahim, I.H., Gilfoyle, L., Reynolds, R., and Voulvoulis, N. (2019). Integrated catchment management for reducing pesticide levels in water: engaging with stakeholders in East Anglia to tackle metaldehyde. *Science of the Total Environment*, *656*, 1436–1447.

National Farmers' Union (NFU) (2018). *NFU Water Bank*. NFU. www.nfuonline.com/cross-sector/environment/water/water-must-read/nfu-water-bank/.

National Farmers' Union (NFU) and Cranfield University (2018). *Assessing Opportunities for Secondary Markets for Water in Response to Proposed Abstraction Reforms: Key Findings*. www.nfuonline.com/nfu-online/science-and-environment/ irrigation-and-abstraction/cranfield-nfu-report-secondary-markets-oct-18/.

Ofwat (2018). *Information Notice – Draft Water Resource Management Plans 2019: Overview of Ofwat's Responses*. www.ofwat.gov.uk/publication/18-12-draft-water -resources-management-plans-2019-overview-ofwats-responses/.

Ofwat (2019). Water trading ('Bulk supplies') register 2019–20. Anglian Water, ANH bulk supply data. https://www.ofwat.gov.uk/water-trading-bulk-supplies-register -2019-20/.

Royal Society for the Protection of Birds (RSPB) (2019). *Catfield Fen*. www.rspb.org .uk/our-work/our-positions-and-casework/casework/cases/catfield-fen/.

Tabas, A., Garrick, D., and Tremolet, S. (2019). Engagement strategies for water resource management in Eastern England. Nature Conservancy internal report (unpublished).

Tinch, R. (2009). *Assessing Socio-economic Benefits of Natura 2000 – a Case Study on the Ecosystem Service Provided by the Sustainable Catchment Management Programme (SCaMP)*. Output of the project Financing Natura 2000, Cost estimate and benefits of Natura 2000 (Contract No.: 070307/2007/484403/MAR/B2), 28 pp. + Annexes.

University of Cambridge and Anglian Water (2013). *Sustainable Water Stewardship: Innovation through Collaboration*. http://www.cisl.cam.ac.uk/business-action/ business-nature/natural-capital-impact-group/pdfs/sustainable-water-stewardship -innovation-through-c.pdf/view.

Water Resources East (WRE) (2020). *Collaborating to Secure Eastern England's Future Water Needs: Our Initial Water Resource Position Statement*. www.wre.org .uk/wp-content/uploads/2020/04/WRE-Initial-statement-of-resource-need-FINAL .pdf.

Water Resources East (WRE) (2021). *Water Resources East: Updated Resource Position Statement*. https://wre.org.uk/wp-content/uploads/2021/03/WRE-RPS -report-March-2021-FINAL.pdf (accessed 24 March 2021).

Wheatley (2021). Products: Wheatley Watersource. https://www.wheatleysolutions.co .uk/products/wheatley-watersource/ (accessed 24 March 2021).

Wheeler, K.G., Robinson, C.J., and Bark, R.H. (2018). Modelling to bridge many boundaries: the Colorado and Murray–Darling River Basins. *Regional Environmental Change*. 10.1007/s10113-018-1304-z.

Wheeler, S.A., Bark, R., Loch, A., and Connor, J. (2015). Agricultural water management. In A. Dinar and K. Schwabe (eds), *Handbook of Water Economics* (pp. 71–86). Edward Elgar Publishing.

Wheeler, S.A., Loch, A., Crase, L., Young, M., and Grafton, R.Q. (2017). Developing a water market readiness assessment framework. *Journal of Hydrology, 552*, 807–820.

12. Assessment of water markets in Chile

Guillermo Donoso, Pilar Barria, Cristian Chadwick and Daniela Rivera

12.1 INTRODUCTION

Chile was an early adopter of water right markets. The Chilean Water Code of 1981 established that water rights are transferable in order to facilitate markets as an allocation mechanism. An enabling factor was Chile's tradition and culture, dating back to colonial times, of managing water resources with water rights. Water right (WR) markets have been documented and are more prevalent in areas of water scarcity. They are driven by demand from relatively high-valued water uses and facilitated by low transaction costs in valleys with flexible water distribution infrastructure and where Water User Associations assist. In the absence of these conditions, trading has been rare and WR markets have not become institutionalized. WR markets in Chile face many challenges in order for them to deliver their full potential as an efficient water allocation mechanism.

The objective of this chapter is to apply the water market readiness assessment (WMRA) framework developed by Wheeler et al. (2017) to detect those issues that must be considered in order establish an effective water allocation mechanism based on a WR market.

12.2 CLIMATE AND HYDROLOGICAL SETTING

Continental Chile is a narrow and long country located between 17.5°S in the north and 55.9°S in the south, on the west slope of the Andes in South America. Chile spans a unique variety of climates, from the desertic north to the very wet rain-oceanic climate in the south, including the central sub-humid Mediterranean climate region.

The Directorate of Water Resources (DGA) of the Ministry of Public Works (MOP) inventory indicates that there are 101 catchments in the Chilean territory (DGA, 2016). Based upon hydroclimatological and topographic characteristics, the DGA has grouped Chilean catchments in four macro-zones:

North (15°S to 25°S), Central (25°S to 40°S), South (40°s to 45°S) and Austral macro-zones, respectively. The large geographical and climatic variability of Chile sets the context for an unequal distribution of runoff, with average values ranging from about 36.9 m³/s in the northern catchments to 20258 m³/s in the Austral catchments.

The DGA monitoring network includes 2895 valid stations currently operating along the national territory. However, as 65 per cent of them are located in the North and Central macro-zones, the South and Austral macro-zones catchments are underrepresented.

Much has been advanced in characterizing and understanding Chilean hydrology. The recently published *Water Atlas* (*Atlas del Agua*; DGA, 2016) offered a general view regarding current hydrology. However, important shortcomings still remain, particularly regarding:

1.   Monitoring and understanding the hydrological processes of high-elevation catchments.
2.   Quantification of the glaciological contribution to snowmelt-dominated runoff catchments.
3.   Monitoring and understanding the hydrology of catchments of the South and Austral macro-zones.
4.   Modelling and quantifying the impacts of climate change on water resources.

Furthermore, McPhee et al. (2017) has stated that in general, the quantification of groundwater balance is very limited, particularly in the North and Central macro-zones due to a poor well network. Acknowledging these limitations, the DGA conducted the National Plan of evaluation of water sources (DGA, 2017a). Based upon their criticality, 40 aquifers were prioritized within the Plan, to implement hydrogeological models to improve the water balance estimate.

Finally, although river–aquifer interaction in Chile is usually very active (Peña, 1992), there is limited information, especially south of Santiago (Arumí Ribera and Oyarzún Lucero, 2006). This has led to an incorrect spatial representation of the hydrogeological processes (Viguier et al., 2018).

Currently, an update of the Chilean Water Balance is being conducted (DGA, 2017b) in order to overcome some of these shortcomings. A significant improvement is the estimation of the groundwater component of the water balance, which was not considered in the previous study. Also, considering the non-stationary nature of the hydrology, the water balance will provide modelled based projections of runoff in the 101 Chilean catchments, considering climate change scenarios (DGA, 2017b).

12.3 LEGAL AND INSTITUTIONAL FRAMEWORK

12.3.1 Water Institutional Framework

The Chilean water institutional framework is characterized by the coexistence of centralized and decentralized institutions (Vergara and Rivera, 2018). Centralized organizations comprise the administrative bodies of the state. Decentralized bodies, on the other hand, are represented by Water User Associations (WUAs), which are not part of the state administration.

Under the 1981 Water Code (WC81), the state reduced its intervention in water resource management to a minimum and increased the management powers of water right holders organized in WUAs. Water delivery is also largely via privatized networks.

12.3.2 Public Institutions

There are 30 administrative organizations with jurisdiction over water issues (World Bank, 2013). The main institution is the DGA, a centralized technical entity with multiple powers and attributions. These include the following:

1. Original water rights (WRs) allocation, constitution or creation, over both surface waters and groundwaters through a concession procedure regulated in the WC81.
2. Develop hydrological plans and provide measures to prevent depletion of aquifers.
3. Conduct water research, measurement and monitoring of both water quality and quantity.
4. Monitor and enforce the use of surface and groundwater.
5. Oversee WUAs.
6. Keep the Public Water Registry (CPA), which is the administrative register containing all information on the country's water resources and WRs.

There are problems that affect the normal and optimal exercise of centralized water management in Chile. According to the World Bank (2011, 2013) the most significant problems and issues are the following:

1. Lack of a complete and updated national water reporting system of the country's water resources and WRs.
2. High fragmentation and dispersion of powers and functions, without effective coordination and collaboration mechanisms among them.
3. Lack of hydrological planning at the national and catchment level.
4. Lack of stakeholder participation in water resource management.

5.   Lack of a stronger institutional presence of the DGA.

Due to this, Chile's public institutions are not considered to be fully respected and trustworthy (World Bank, 2013; Vergara, 2014). Additionally, this situation has led to significant increase in conflicts between the WR holders, stakeholders and the DGA (Guerra, 2016; Rivera et al., 2016; Rivera Bravo, 2015; Herrera et al., 2019).

12.3.3   Water User Associations

The WC81 establishes that WR holders, organized in collective WUAs, are responsible for water management at the local level (Vergara et al., 2013). WUAs have existed in Chilean water resource management since colonial times, despite the fact that the first law that regulates them was passed in 1908 (Obando, 2009). Three types of WUAs exist in Chile: small-scale water communities (CAs), canal user associations (ACs), and vigilance committees (JdVs). A CA is any formal group of users that share a common superficial or groundwater source. There are more surface water communities than groundwater communities. ACs have jurisdiction over common surface water infrastructure. JdVs act at the level of a natural source, and their members include CAs, ACs and individual WR holders. They are responsible for the joint management and administration of a basin's waters, including both surface water and groundwater. Nevertheless, their actions have focused on surface waters. WUAs also maintain their own water user registries, which are often more complete and up to date than official registries (Hearne and Donoso, 2014).

   Many of these WUAs have professional management. However, a significant proportion of these have not updated their capacity to meet actual water management challenges (Montginoul et al., 2016; DGA, 2018; Rinaudo and Donoso, 2019). Due to these concerns, the DGA, National Irrigation Committee (CNR), and Hydraulic Public Works Directorate (DOH) have implemented programmes to strengthen WUAs (DGA, 2018).

12.3.4   Water Rights Legislation

The first legal text on waters issued in Chile in 1819 established the public nature of water (Rivera, 2013). Other regulatory texts contain implicit or indirect declarations with similar orientations, but the first explicit references are found in the Civil Code of 1857, and the Water Codes of 1951 and 1967.

   At present, the public nature of waters available in natural sources is formulated explicitly in article 595 of the Civil Code and article 5 of the WC81; and implicitly in the Political Constitution (art. 19 No. 24 final paragraph, in relation to art. 19 No. 23 of the Political Constitution). Thus, any use of water requires a WR legal title constituted by the competent authority in a procedure expressly regulated in the WC81 (Rivera, 2013).

The WC81 recognizes the following types of WRs, in terms of their origin:

1. Ancient rights, which consist of two types:
   a. Those constituted in accordance with the formalities, terms and procedures established by regulations prior to the WC81.
   b. Those recognized based on customary water uses. These are de facto practices exercised for many years, outside the constitutive procedures established by law. Since the promulgation of the WC81, efforts have been made to regularize and register customary WR. However, estimates of customary WR that are not registered range from 60 to 90 per cent. This can be explained by the fact that courts have protected unregistered rights, and thus undermined the registration requirement.
2. New rights, which refer to all titles constituted in accordance with the concession procedure regulated in the WC81.

All of these WRs, ancient and new, constituted and recognized, have the same legal value and protection.

The WC81 specifies consumptive WRs for both surface water and groundwater, and non-consumptive WRs for surface waters. Non-consumptive WRs allow the user to divert water from a river with the obligation to return the same water unaltered to its original water source. Consumptive WRs do not require that water be returned once it has been used.

Additionally, consumptive and non-consumptive WRs can be permanent or contingent, and exercised continuously, discontinuously or alternating. Permanent WRs correspond to runoff with a probability of exceedance greater than 85 per cent. Contingent WRs correspond to streamflow which can be used only after permanent WRs have been satisfied, and respecting stipulated ecological river runoff. Ecological runoff is defined as the minimum runoff required to satisfy the environmental needs of the river ecosystem. It is important to note that minimum ecological flows have not been enforced.

A permanent WR combines volumetric caps per unit in times of plenty, with shares in times of scarcity. This has proven to be appropriate (World Bank, 2011).[1] Continuous rights are WRs that allow users to extract water continuously over time. On the other hand, discontinuous rights are those that only permit water to be used at given time-periods. Finally, alternating rights are those in which water is distributed among two or more persons who use the water successively.

The WC81 established that WRs are independent of land and are transferable, in order to enable markets as an allocation mechanism. Furthermore, the WC81 allows for freedom in the use of water to which an agent has a WR;

thus, WRs are not sector-specific. Additionally, the WC81 abolished water preferential lists, present in the previous Water Codes of 1951 and 1967. Finally, WRs do not expire, and do not consider a 'use it or lose it' clause.[2]

12.3.5 Water Resources Information System

Water resources information generation management system and registry
The DGA is responsible for maintaining and updating the Water Resources Information Registry, which is a fundamental tool for water management and WR markets.

Current maintenance of water resource information
There is no protocol or norm on the periodicity of information maintenance. However, the system is periodically updated, a few times a year. An important number of stations provide data in real time. However, others are updated two to three times per year.

All granted WRs must be recorded as public deeds and registered in the Water Property Registry of the corresponding Real Estate Registrar (CBR). Additionally, all WRs granted under the WC81 must also be listed in the CPA.

Identification of information gaps
There are serious shortcomings regarding the quality and availability of critical data such as WRs. Firstly, under existing water legislation, holders of WRs granted prior to the WC81 are not required to inform the DGA of their existing entitlements, or of WR transfers. On the other hand, ancient customary WRs are not recorded or listed in any registry. Thus, the number of WRs that are registered and recorded is notoriously lower than that of existing WRs.

Secondly, even though WR holders have the obligation to implement measurement systems, as well as to store and report this information to the DGA, this requirement has not been imposed. Thus, the DGA's public database does not contain accurate information on effective extraction and use of water resources (World Bank, 2011). These problems affect the definition of water availability and thus the WR granting procedure, leading to over-allocation enhancing water conflicts.

Finally, there are many WRs registered as 'water shares', meaning that their volume equivalents vary depending on the sub-catchment and local WUAs, a factor that further complicates the potential to quantify actual water allocations. A recent study updated and systematized WUA information that was scattered in different institutions outside the DGA (DGA, 2018). The study improved the WR Registry of the DGA, including several new WRs. It also provided some estimations of the equivalence of water shares in litres per second (l/s) of 22 different WUA.

12.3.6   Mechanisms to Ensure Sustainable Water Exploitation

The mechanism to ensure sustainable water exploitation is through the definition of a sustainable exploitation flow (SEF), which ensures the long-term equilibrium in the hydrological system (DGA, 2008). The SEF must not generate negative impacts on the environment. The first problem of the mechanism is the unspecified length of 'long-term', because considering the non-stationary nature of hydrogeological processes, the length of the period considered for the evaluation can significantly change the SEF. Moreover, a critical drawback is that in order to accomplish the target of the SEF, detailed and robust hydrogeological models which simulate the response of the aquifer are needed, but are not available. Given the uncertainties in determining the total amount of sustainable water extractions, decree 203 issued in 2013 establishes specific thresholds for interactions between different points of groundwater extractions that should lead to water extraction restrictions or prohibitions.

The WC81 considers three measures to limit groundwater extraction: a temporary reduction in the exercise of WRs, the declaration of a restriction area and of a prohibition zone. None of these measures exist with respect to surface waters. A temporary reduction in the exercise of WRs may be applied when groundwater extraction affects the sustainability of the aquifer or causes damage to other WR holders. The possibility of limiting withdrawals has been contemplated since 1983. However, this restriction has never been implemented, despite groundwater sustainability concerns.

A restriction area is declared when there is risk of severe reduction of water available in the aquifer, with consequent damage to constituted or recognized WRs. The restriction area must be declared by the DGA, at its own initiative or at the request of a groundwater right holder. A restriction area does not imply a total closure of the groundwater source, as the DGA has the possibility to grant provisional rights. These rights have a more precarious status than definitive rights, as they can be annulled by the DGA. On the other hand, provisional groundwater rights can become permanent if they are used continuously for at least five years, on condition that they do not affect other groundwater right holders.[3] The provisional WRs are restricted to be less than 25 per cent of permanent water rights.

The declaration of a prohibition area arises from the need for greater protection of the aquifer (Rivera Bravo, 2015; Rivera 2018), and hence no further rights can be granted. Additionally, a groundwater community among all groundwater right holders must be formed when a restriction or prohibition area has been declared. Although at least 159 groundwater communities must exist, only 13 have in fact been established.

The DGA may also, at the request of the respective JdV, declare surface water sources to be exhausted, restricting the constitution of new permanent consumptive WRs. Finally, WUAs have broad attributions to protect the sustainable use of a water resource.

12.3.7   Monitoring and Enforcement Regulation

WUAs are responsible for monitoring and enforcing water extraction. However, only a reduced number have fulfilled this obligation (DGA, 2018). Considering this, the WC81 reform (approved in 2005), established new legal powers for the DGA to strengthen its monitoring and enforcement capabilities. However, the monitoring conducted by the DGA led to only seven official sanctions in a 12-year period. This is due to the fact that the DGA does not have the necessary resources (human, technical and financial) to monitor all extractions.

Additionally, Estévez (2017) points out that most of the sanctions for illegal water extractions had penalties of about US$1450. Thus, expected non-compliance costs have been lower than the economic benefit of not complying. In fact, some farmers openly acknowledge that they do not comply with the terms of their WRs (Rinaudo and Donoso, 2019).

Some of these deficiencies have been addressed by law N° 21,064 of 2018, that modifies inspections and sanctions. The fines of the new sanctions range from approximately US$725 to US$144840. The new law clearly empowers the DGA. However, its approval did not include an increase in the DGA's budget and personnel. Thus, the DGA is still short of human and economic resources to properly fulfil its duty.

12.4   REALLOCATION OF WATER USE RIGHTS THROUGH MARKETS REGULATION

12.4.1   Performance

Although market reallocation of water has not been common throughout most of Chile, the existence of consumptive WR (CWR) markets has been documented. Hearne (2018) points out that there has been active trading for CWRs in basins where water is scarce with a high economic value. The presence of flexible gates in the distribution infrastructure[4] as well as well-managed WUAs have reduced transaction costs and fostered CWR transactions. Many CWR transactions have been for relatively small amounts of water and for low transaction amounts. This indicates that transaction costs have often not been prohibitive.

The majority of CWR transactions have been between agricultural users. In general, intersectoral water transfers have not been frequent. Even in active markets of the Limarí basin, only 2 per cent of transactions between 2000 and 2016 transferred water out of agriculture (Hearne, 2018). In the Copiapó Basin, where mining and high-valued agriculture have competed for increasingly scarce water, CWR markets have been active between farmers, with prices rising with increased scarcity (Donoso et al., 2014a).

Prices have been highly variable. This large price dispersion is due to the lack of reliable public information on WR prices and transactions (Donoso, 2015). Given this, each WR transaction is the result of a bilateral negotiation between an interested buyer and seller of a WR in which each agent's information, market experience and negotiating capacity is important in determining the final result (Donoso et al., 2014b). The market for non-consumptive WRs (NCWR) has traded approximately 2.6 million l/s between 2009 and 2014 (Cristi et al., 2014). In terms of market activity, measured as the percentage of the total volume of granted NCWRs, about 20 per cent of granted NCWRs have been traded (Cristi et al., 2014).

12.4.2 Externalities

Return flows

Many Chilean basins are characterized by return flows, which imply that water flows are consecutively used by different users throughout the basin. Peña (2018) shows, for example, that water in the Coquimbo region is used between two and four consecutive times. In the North macro-region, this implies that 88 per cent of effective runoff is used before it reaches the sea (Peña, 2016).

This has led to externalities, since river basins located in the North and Central macro-regions have historically been managed by river sections, considering that these sections are independent of each other. In this system, WRs have been granted for each section, with no obligation to conserve flows for downstream sections. Thus, any changes in water use efficiencies, due to the adoption of efficient irrigation technologies or WR transfers to more water-intense users, leads to reduced water flows in downstream sections. This externality is not considered in current legal regulations.

Groundwater allocation based on foreseeable use

The sustainability of groundwater is at risk due to the over-allocation of WRs on the basis of a foreseeable use factor. Prior to 2000, the DGA estimated available water flow that could be allocated based on an estimate of its intended foreseeable use. For example, since the majority of irrigated agriculture irrigates only in late spring and late summer, the DGA assumed that its WRs would in reality represent 20 per cent of the total requested use. The theoretical use factor for drinking water and the mining industry was 75 per cent.

The problem is that the granted groundwater right does not specify this foreseeable use, and thus allows for the extraction of 31500 m³/year. Thus, if a WR is transferred from agriculture to mining or domestic water supply, the volume of extracted groundwater increases significantly. For example, Rinaudo and Donoso (2019) estimated that granted volume increased 76 per cent in the Copiapó aquifer, due to increases in irrigation efficiency that translated into an increase in irrigated surface, and to WR transfers from agriculture to mining.

12.5 WATER MARKET READINESS ASSESSMENT (WMRA) APPLICATION

This section applies the WMRA framework (Wheeler et al., 2017) to examine the appropriateness of Chile's water right markets as an allocation mechanism, and to identify how they might be improved to facilitate low-cost trading.

12.5.1 Property Rights/Institutions

Notwithstanding the progress and positive aspects of Chile's institutional framework, robustness, transparency and security are not features that have particularly been achieved. In fact, the fragmentation of powers and functions among different public and private entities, and the lack of coordination among them, has impeded the construction of an effective, transparent and secure institutional framework.

The WC81 maintained water as national property for public use and granted tradeable WRs that are legally separate from land property, so as to facilitate water reallocation through WR markets. Besides constituted WRs, the WC81 recognizes WRs previously constituted and WRs based on customary water uses. All of these WRs, ancient and new, constituted and recognized, have the same legal value and protection. WUAs and the DGA must monitor and enforce water use. However, as previously pointed out, the enforcement regime has not been effective and efficient. Hence, there is little incentive for WR holders to comply.

12.5.2 Hydrology

Currently, the Water Resources Information Registry presents problems, specifically related to data limitations. The DGA's public database does not contain accurate information on effective extraction and use of water resources. These problems affect the definition of water availability and have led to over-allocation enhancing water conflicts. Poorly distributed and short records of fluviometric and meteorological stations have restricted water planning, which result in poor estimations of the water balance and thus in the determination of water availability. This has particularly hampered estimations of future water availability. To date, model-based projections of runoff are not considered by DGA for water planning, which could exacerbate water scarcity in the future.

Currently, the water system in Chile does not properly consider the interception of runoff. An additional shortcoming is the inaccuracy in the delimitation of aquifers, which is fundamental to understand the recharge and surface–aquifer interactions. Thus, groundwater interaction with surface water is not fully understood, documented and monitored, especially in the Central and South macro-zones. Currently, an update of the Chilean Water Balance is being conducted, which seeks to overcome some of the shortcomings of poor estimations of the water balance and thus in the determination of water availability.

12.5.3   System Type

The state has been involved in the development of large irrigation infrastructure under the build, operate and transfer (BOT) model. Due to growing water scarcity and projected climate change impacts on water resources, Chile's National Water Resources Policy considers a significant increase in the investment of large reservoirs. Chile does not have long-distance infrastructure that could interconnect the more than 100 basins. Therefore, water markets in Chile are local and within river sections.

Historically, the private sector developed water infrastructure, and today much of the nation's canal infrastructure is operated by WUAs. Well-managed WUAs have facilitated WR transactions and have been effective in reducing transaction costs. In basins with rigid water distribution infrastructure, WR buyers incur significant costs since fixed canal dividers must be modified. These costs are present throughout Chile, with the exception of the few basins with flexible adjustable canal gates, and have reduced trading. The WC81 allows for permanent and temporary trades to take place. However, active temporary trades are present only in basins with flexible water distribution infrastructure.

12.5.4   Externalities and Governance

There is little regulation of market trades, and traded consumptive WRs are not required to maintain return flows. To reduce potential negative effects on third parties and/or the environment, when a transfer of a WR implies a change of water intake location, it must be authorized by the DGA. Additionally, the Environmental Impact Assessment System requires water users to mitigate or compensate environmental damages that may result from the transfer of WRs. However, WR transfers to more water-intensive users leads to reduced return flows in downstream sections. This externality is not considered in current legal regulations. WR markets have also exacerbated the unsustainability of groundwater management, due to an over-allocation of WRs generated by an inadequate specification.

There is little concern about unused consumptive rights for surface water (Donoso, 2015). On the other hand, non-use of non-consumptive WRs is an issue. To reduce the non-use of WRs, the 2005 WC81 reform introduced a non-use tariff, which has been shown to be effective (Valenzuela et al., 2013). Non-use of WRs is expected to continue to reduce in the future due to the projected increase in the non-use tariff. However, as groundwater levels fall due to over-extraction, the number of unused groundwater rights will increase due to drying wells.

12.5.5   Adjustment

Water scarcity triggered the initiation of water markets in Chile. The market transfer of WRs has produced significant economic gains from trade for both buyers and sellers, particularly in basins located in the North and Central macro-regions, characterized by higher water scarcity. These economic gains drive intersectoral trades and WR trades between farmers. However, WR trading has not been accepted by all actors. Several farmers maintain that water and land should not be separated. This has kept many farmers from offering WRs for sale without also selling the corresponding land. Additionally, WR trading does not have full political acceptability. For example, some members of Congress proposed a water code reform stating that water management based on productive and market logic is dysfunctional to the protection, rational use and equitable distribution of water resources.

Legislative changes have been hindered by constraints of the 1980 Constitution and the lack of significant political support. A major reform occurred in 2005 after 13 years of political debate in Congress. Since 2012, Congress has been debating a new water code reform. The most relevant proposals are to prioritize human water consumption, environmental protection, establish expiration dates on newly granted WRs, and to introduce a 'use it or lose it' clause. Social movements pushing for reform have emerged due to the lack of political decision.

12.5.6 Entitlement Registers and Accounting Systems

WR information is scattered among different organizations, thus a considerable number of WRs have not been registered in the DGA database. Ancient customary WRs are not recorded or listed in any registry. Thus, the number of WRs that are registered and recorded is notoriously lower than the actual number of WRs. Hence, Chile's water resources and Water Resources Information Registry is not reliable and trustworthy. Additionally, given that only formal registered WRs can be traded, the fact that most rights remain unregistered hinders the effectiveness of Chile's WR market.

Chile's WR market developed mostly in the absence of market intermediaries. It is characterized by large price dispersion for homogeneous WRs. This large price dispersion is due, in great part, to the lack of reliable public information on WR prices and transactions. The lack of a price-revealing mechanism has significantly hampered the effectiveness of the WR market as a reallocation mechanism.

12.6 SUMMARY

The above analysis indicates that, although weaknesses exist, Chile has significantly advanced in achieving Trade Step 2 of the WMRA framework. The main elements that have hindered WR market effectiveness in Chile are the lack of: (1) WR and WR market information; (2) regularization of customary WRs; (3) WR and water use enforcement; and (4) externalities modelling and understanding.

NOTES

1. Groundwater rights have not been exercised as shares, as this requires an approval of the groundwater community, which has not occurred even when aquifers are in risk of depletion.
2. At present, these two characteristics are highly questioned (González, 2018).
3. No provisional groundwater rights have become permanent, even though they have been in use for more than five years.
4. A large percentage of the water distribution infrastructure divides water among users of a channel by fixed gates, which is a rigid system of water distribution that is costly to modify.

REFERENCES

Arumí Ribera, J., and Oyarzún Lucero, R. (2006). Las aguas subterráneas en Chile. *Boletín Geológico y Minero, 117*(1), 37–45.

Cristi, O., Donoso, G., and Melo, O. (2014). *Análisis Estimación del Precio Privado de los Derechos de Aprovechamiento De Aguas*. Comisión Nacional de Riego, Ministerio de Agricultura. Santiago, Chile.

DGA (2008). *Manual de Normas y Procedimientos Para la Administración de Recursos Hídricos*. http://documentos.dga.cl/ADM5016.pdf.

DGA (2016). *Atlas del Agua, Chile 2016*. Dirección General de Aguas, Ministerio de Obras Públicas, Santiago, Chile. http://sad.dga.cl/ipac20/ipac.jsp?session =158A05U86851U.6830238&profile=cirh&source=~!biblioteca&view= subscriptionsummary&uri=full=3100001~!5891~!0&ri=1&aspect=subtab39& menu=search&ipp=20&spp=20&staffonly=&term=atlas+de+agua&index=.GW& uindex=&aspect=subtab39&menu=search&ri=1.

DGA (2017a). *Inventario Nacional de Acuíferos*. Dirección General de Aguas, Ministerio de Obras Públicas, Santiago, Chile. http://sad.dga.cl/ipac20/ipac.jsp ?session=1S37K301U5187.1395208&profile=cirh&uri=full%3D3100001%7E %215993%7E%210&booklistformat=.

DGA (2017b). *Actualización del Balance Hídrico Nacional*. Dirección General de Aguas, Ministerio de Obras Públicas, Santiago, Chile. https://cambioglobal.uc.cl/ images/proyectos/MetodologiaBalanceHidrico_DGA_CCG-UC.pdf.

DGA (2018). *Diagnóstico Nacional de Organizaciones de Usuarios*. Dirección General de Aguas, Ministerio de Obras Públicas, Santiago, Chile. http://sad.dga.cl/ipac20/ ipac.jsp?session=158505507Q60V.6830556&profile=cirh&source=~!biblioteca& view=subscriptionsummary&uri=full=3100001~!6056~!3&ri=1&aspect=subtab39 &menu=search&ipp=20&spp=20&staffonly=&term=%22marzo+03/19%22&index =.GW&uindex=&aspect=subtab39&menu=search&ri=1.

Donoso, G. (2015). Chilean water rights markets as a water allocation mechanism. In M. Lago, J. Mysiak, C. Gómez, G. Delacámara and A. Maziotis (eds), *Use of Economic Instruments in Water Policy: Insights from International Experience* (pp. 265–278). Springer. https://doi.org/10.1007/978-3-319-18287-2.

Donoso, G., Blanco, E., Vergara, A., and Rivera, D. (2014a). *Capacitación y apoyo a comunidades de aguas subterráneas en el valle de Copiapó, Región de Atacama*. Comisión Nacional de Riego, Ministerio de Agricultura. Santiago, Chile.

Donoso, G., Melo, O., and Jordán, C. (2014b). Estimating water rights demand and supply: are non-market factors important? *Water Resource Management*, *28*(12), 4201–4218. https://doi.org/10.1007/s11269-014-0739-3.

Estévez, C. (2017). *Proyecto de Ley que Modifica el Código de Aguas Boletín 8149-09*. https:// www.camara.cl/pdf.aspx?prmID=115758&prmTIPO=DOCUMENTOCOMISION.

González, P. (2018). *Reforma el Código de Aguas: Ley, Tramitación y Principales Modificaciones*. Asesoría Técnica Parlamentaria, Biblioteca del Congreso Nacional. http://www.senado.cl/appsenado/index.php?mo=tramitacion&ac=getDocto& iddocto=4354&tipodoc=docto_comision.

Guerra, F. (2016). Resolución de disputas en el contexto de los conflictos en torno al agua en Chile: Una respuesta institucional múltiple. *Revista de Derecho Ambiental*, *4*(6), 205–223. https://revistaderechoeconomico.uchile.cl/index.php/RDA/article/ view/43322.

Hearne, R. (2018). Water markets. In G. Donoso (ed.), *Water Policy in Chile* (pp. 117–130). Springer. https://doi.org/10.1007%2F978-3-319-76702-4.

Hearne, R., and Donoso, G. (2014). Water markets in Chile: are they meeting needs? In W. Easter and Q. Huang (eds), *Water Markets for the 21st Century: What Have We Learnt?* Issues in Water Policy 11. Springer-Verlag. https://doi.org/10.1007/978 -94-017-9081-9__6.

Herrera, M., Candia, C., Rivera, D., Aitken, D., Brieba, D., et al. (2019). Understanding water disputes in Chile with text and data mining tools. *Water International, 44*(3), 302–320. https://doi.org/10.1080/02508060.2019.1599774.

McPhee, J., Mengual, S., and Macdonell, S. (2017). A modelling study of the seasonal snowpack energy balance at three sites along the Andes Cordillera: regional climate and local effects. *Geophysical Research Abstracts, 19*, EGU2017-11019.

Montginoul, M., Rinaudo, J., Brozović, N., and Donoso, G. (2016). Controlling groundwater exploitation through economic instruments: current practices, challenges and innovative approaches. In A.J. Jakeman, O. Barreteau, R.J. Hunt, J.-D. Rinaudo and A. Ross (eds), *Integrated Groundwater Management* (pp. 551–581). Springer.

Obando, I. (2009). Estructura y jurisdicción de las organizaciones de usuarios de aguas en chile durante el siglo XIX. *Revista de Derecho de la Pontificia Universidad Católica de Valparaíso, 32*, 107–132. https://doi.org/10.4067/S0718-68512009000100002.

Peña, H. (1992). Caracterización de la calidad de las aguas naturales y contaminación agrícola en Chile. In FAO (ed.), *Prevención de la contaminación del agua por la agricultura y actividades afines: anales de la Consulta de Expertos organizada por la FAO* (pp. 75–86). FAO, ONU.

Peña, H. (2016). *Desafíos de la seguridad hídrica en América Latina y el Caribe.* Serie Recursos Naturales e Infraestructura No. 178, CEPAL. https://repositorio.cepal.org/handle/11362/40074.

Peña, H. (2018). Integrated water resources management in Chile: advances and challenges. In G. Donoso (ed.), *Water Policy in Chile.* Global Issues in Water Policy, Vol. 21 (pp. 197–207). Springer.

Rinaudo, J.-D., and Donoso, G. (2019). State, market or community failure? Untangling the determinants of groundwater depletion in Copiapó (Chile). *International Journal of Water Resources Development, 35*(2), 283–304. https://doi.org/10.1080/07900627.2017.1417116.

Rivera, D. (2013). *Usos y derechos consuetudinarios de aguas: su reconocimiento, subsistencia y ajuste.* LegalPublishing, Thomson Reuters.

Rivera, D. (2018). Alumbrando conflictos: disponibilidad y asignación de derechos de aguas subterráneas en la jurisprudencia chilena. *Revista de Derecho, 31*(1), 159–183. https://doi.org/10.4067.

Rivera, D., Godoy-Faúndez, A., Lillo, M., Alvez, A., Delgado, V., et al. (2016). Legal disputes as a proxy for regional conflicts over water rights in Chile. *Journal of Hydrology, 535*, 36–45.

Rivera Bravo, D. (2015). Diagnóstico jurídico de las aguas subterráneas. *Ius et Praxis, 21*(2), 225–266. https://doi.org/10.4067/s0718-00122015000200007.

Valenzuela, C., Fuster, R., and Leon, A. (2013). Chile: ¿ Es eficaz la patente por no uso de derechos de aguas? *Revista Cepal, 109*, 175–198. https://repositorio.cepal.org/bitstream/handle/11362/11578/109175198_es.pdf?sequence=1&isAllowed=y.

Vergara, A. (2014). *Crisis institucional del agua: Descripción del modelo jurídico, crítica a la burocracia y necesidad de tribunales especiales.* Legal Publishing, Thomson Reuters.

Vergara, A., and Rivera, D. (2018). Legal and institutional framework of water resources. In G. Donoso (ed.), *Water Policy in Chile.* Global Issues in Water Policy, Vol. 21 (pp. 67–85). Springer.

Vergara, A., Donoso, G., Rivera, D., Blanco, E., and Moyano, V. (2013). Aguas y energía: propuestas para su autogobierno y resolución especializada de conflictos. In I. Irrarazabal, M. Morandé and M. Letelier (eds), *Propuestas para Chile*

2013 (pp. 241–270). Centro Políticas Públicas – UC. https://politicaspublicas .uc.cl/publicacion/concurso-de-politicas-publicas-2/propuestas-para-chile-2013/ propuestas-para-chile-2013-capitulo-viii-aguas-y-energia-propuestas-para-su -autogobierno-y-resolucion-especializada-de-conflictos/.

Viguier, B., Jourde, H., Yáñez, G., Lira, E., Leonardi, V., et al. (2018). Multidisciplinary study for the assessment of the geometry, boundaries and preferential recharge zones of an overexploited aquifer in the Atacama Desert (Pampa del Tamarugal, Northern Chile). *Journal of South American Earth Sciences, 86*, 366–383.

Wheeler, S.A., Loch, A., Crase, L., Young, M., and Grafton R.Q. (2017). Developing a water market readiness assessment framework. *Journal of Hydrology, 552*, 807–820. https://doi.org/10.1016/j.jhydrol.2017.07.010.

World Bank (2011). *Diagnóstico de la gestión de los recursos hídricos*. Departamento del Medio Ambiente y Desarrollo Sostenible, Santiago, Chile, Dirección General de Aguas, Ministerio de Obras Públicas. https://doi.org/10.1029/2001WR000748.

World Bank (2013). *Estudio para el mejoramiento del marco institucional para la gestión del agua*. Santiago, Chile, Dirección General de Aguas, Ministerio de Obras Públicas de Chile. http://sad.dga.cl/ipac20/ipac.jsp?session=1N85056062P4Y .6831867&menu=search&aspect=subtab39&npp=10&ipp=20&spp=20&profile= cirh&ri=&term=Estudio+para+el+mejoramiento+del+marco+institucional+para+ la+gesti%C3%B3n+del+agua&index=.GW&x=0&y=0&aspect=subtab39.

13. Ready or not? Learning from 30 years of experimentation with environmental water markets in the Columbia Basin (USA)

Gina Gilson and Dustin Garrick

13.1 INTRODUCTION

The Columbia River is often seen as a story of salmon and hydropower, emblematic of the abundant water resources of the Pacific Northwest. But upstream of the mainstem are semi-arid tributaries, which experience additional scarcity during the late summer when irrigation water use coincides with the natural low flows in the hydrograph. Water stress is driven by land use change, urban and exurban growth in semi-arid parts of the basin where groundwater pumping impairs flows and affects both senior water users and endangered fish (Garrick et al., 2009; Settre and Wheeler, 2017). Water markets have brought these stories together by reallocating water from farmers to restore habitat, using hydropower revenues and mitigation funding from residential developers.

This chapter applies the water markets readiness assessment (WMRA) framework (Wheeler et al., 2017) to the Columbia River Basin, one of the world's most advanced laboratories for market-based approaches to water reallocation. Experimentation began in the late 1980s amidst growing interest in free-market environmentalism (Anderson and Leal, 1991), after prior experience with court decisions and administrative decrees proved inadequate to restore stream flows for fish. Policy reforms in the late 1980s and early 1990s enabled reallocation for environmental flows by recognizing environmental purposes as a 'beneficial use' for water, authorizing the transfer of water from irrigation to instream use and mobilizing financing to enter the market on behalf of the environment (Garrick et al., 2009; Garrick et al., 2013). The Columbia Basin illuminates how the factors and institutions associated with water market readiness have developed over time and across different catchments and jurisdictions.

The WRMA envisions three steps – an assessment of hydrological and institutional needs, a market evaluation, and the continuous review and assessment of future needs – progressing either in sequence or in parallel. The framework distils insights for water managers and planners to identify possible barriers to water markets. While barriers to transactions have been analysed in the western United States since the early 1990s (Colby, 1990; Thompson, 1993), research in the last 20 years has specifically focused on drivers and barriers to environmental water transactions in the Columbia River Basin (Garrick et al., 2009; Garrick et al., 2011). In this chapter, we will review the results of our 2017 survey on barriers to transactions and discuss its implications for the WRMA framework and concepts of market readiness.

First, we provide an overview of Steps 1, 2 and 3 of the WMRA framework through reviewing the background context, policy evolution and experience with environmental water transactions in the Columbia Basin. We then discuss our methodology and compare the conceptual framework of our survey to the WMRA. Finally, we discuss the results of our survey and explore the emerging themes in the context of market readiness. Our analysis identifies the barriers to establish these enabling conditions in a mature case study. We demonstrate the importance of parallel, rather than sequenced, investments in policy reform and implementation to address these barriers and adapt. These investments must be tailored to local conditions, although progress across multiple layers of governance is needed to address administrative and financing hurdles.

13.2 THE COLUMBIA BASIN'S BIOPHYSICAL CHARACTERISTICS AND HISTORY

The Columbia River (Figure 13.1) flows 2000 km through the 674 000 km² river basin, covering seven US states (85 per cent of the basin area) and the Canadian province of British Columbia (15 per cent of the basin area). Washington, Oregon, Idaho, and Montana contain the majority of the basin within the United States (US), with small portions in Nevada, Utah and Wyoming. The average annual flow of 237 billion m³ is highly variable in both space and time. Spring snowmelt from mountain streams is a major contributor to most tributaries, with 60 per cent of runoff occurring in late spring and early summer (Kirschbaum and Lettenmaier, 1997), creating the need for irrigation diversions and storage solutions. The basin encompasses a wide range of climatic conditions, from moist maritime in the west to arid conditions in the southeast. Mean annual precipitation across the region is divide into three groups: arid and semi-arid in the Upper Snake and central basin (0–750 mm); humid temperate in the northern headwaters and eastern Cascades (1500–2500 mm); and temperate (750–1500 mm) in the remainder of the basin (Matheussen et al., 2000).

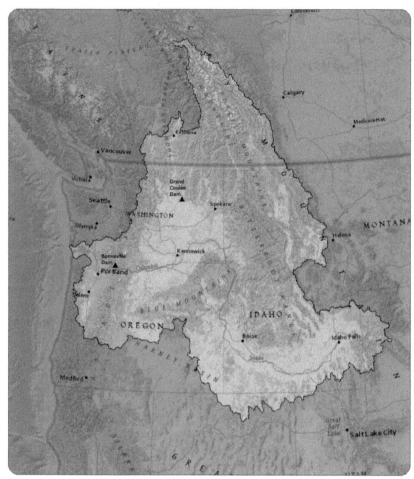

Source: US Environmental Protection Agency.

Figure 13.1   Map of the Columbia River Basin

Pacific salmon have adapted to conditions in the Columbia Basin over a 10 million year history and are integral to the region's cultural history (Cosens et al., 2018). Significant alteration of the Columbia's biophysical system began when European settlers arrived in the 19th century (Hand et al., 2018), with over-fishing spurring the development of salmon hatcheries as early as 1866 (Cosens et al., 2018). Hydropower development drove the decline of salmon and steelhead stocks, with 13 currently listed under the Endangered Species

Act (ESA) (Hand et al., 2018). Today, the Columbia River is considered one of the world's great hydropower rivers, with 14 dams on the main stem and numerous others on its tributaries (Hand et al., 2018), annually averaging 14 000 megawatts of electricity in the US alone (NPCC, 2014). Endangered fish compete with other interests including irrigation, extractive industries and growing cities. The human population in the region continues to increase, in part due to amenity migrants, defined as those who seek a better quality of life by moving towards rural areas that are characterized by natural and/or cultural abundance (Hurley, 2013).

The Northwest Power and Conservation Council (NPCC) has coordinated local recovery efforts within 'ecoprovinces', which are comprised of adjoining groups of ecologically related sub-basins with similar geology, hydrology and climate. These 11 provinces, also called ecoregions, include the entire Columbia River Basin in the US and cover approximately 570 000 km² in Washington, Oregon, Idaho and Montana. These are: Columbia Estuary, Lower Columbia, Columbia Gorge, Columbia Plateau, Columbia Cascade, Inter-Mountain, Mountain Columbia, Blue Mountain, Mountain Snake, Middle Snake and Upper Snake. The 62 sub-basins in the Columbia Basin coincide often, but not always, with the administrative boundaries for watershed planning and water rights administration. Major tributaries include the Snake, Wenatchee, Spokane, Yakima, Kootenai, Deschutes and Willamette rivers (McCoy et al., 2018).

13.2.1   Policy Evolution for Water Trading in the Columbia Basin

The legal basis of water rights in the Columbia Basin is found in the doctrine of prior appropriation, or 'first in time as first in right'. The first appropriator of water is considered 'senior' to later arrivals and is the last to lose access during shortages. The appropriators must maintain beneficial use of the resource without prolonged interruption, and changes to rights cannot impose harm or injury on other upstream or downstream users (Garrick et al., 2009). Under prior appropriation, a cap on water extraction is accomplished implicitly through the physical limits of the river and the finite number of reliable, high-security rights (Aylward, 2008). Basin-wide closures in the Columbia began through reforms and adjudication as early as the 1950s, when minimum water needs were assessed for vulnerable fisheries. States began to recognize the environment as a legitimate water user under common law doctrines and statutory measures that authorized fish and wildlife as eligible beneficial uses (Neuman et al., 2006).

In the Columbia Basin, three aspects of policy reforms were adopted at varying spatial and temporal scales and to varying degrees across the basin. These included the legitimization of environmental water uses under beneficial use doctrine, the enabling of trading between existing rights and new environmental rights, and the authorization and development of funding and organizational capacity to effect transfers for environmental restoration (Garrick et al., 2009).

The first set of reforms authorized wildlife, fish and recreation as beneficial uses for prior appropriation rights. Legal recognition occurred through statutory reform in general water use statutes and legislation targeted at freshwater conservation goals (Neuman et al., 2006). Primary legislative preconditions developed unevenly, first implemented in the early 1970s in Washington and Montana (Garrick et al., 2009). Idaho targeted streamflow protection for fish and wildlife in 1978 through the Minimum Stream Flow Act, and Oregon followed suit in 1987 through the Instream Water Rights Act (Charney, 2005; Loehman and Charney, 2011). The second set of reforms enabled high-priority rights to be leased or transferred instream without losing their underlying priority in Oregon (1987), Montana (1989) and Washington (1989) (Neuman and Chapman, 1999). Reforms raised questions related to governance and who had the authority to acquire, hold and manage water rights. While Montana allowed rights to be held by private entities, Idaho, Oregon and Washington required them to be held by state regulatory agencies (Loehman and Charney, 2011).

Public–private partnerships began to develop between conservation buyers, regulators, funders and sellers to create the capacity for transfers within existing frameworks and allow non-profits to leverage public programmes for conservation work. However, administrative scrutiny and accountability measures created an institutional capacity burden (Garrick et al., 2009). Contract types and transfer mechanisms primarily included permanent acquisitions, temporary leases and the reallocation of water conserved through irrigation efficiency projects (Hardner and Gullison, 2007). An increasing focus on 'scaling up' emphasized the institutionalization of market-oriented transfer mechanisms – such as reverse auctions, water banking, regulated storage, and mitigation for impacts of groundwater pumping on surface flows – while forbearance agreements became a popular tool to temporarily or permanently reduce water use through contractual agreements with private landowners (Garrick et al., 2009; Garrick and Aylward, 2012).

13.2.2 Development of a Water Transactions Programme

State-level policy reforms and watershed-based planning efforts drove environmental transactions from an institutional level. The Northwest Power Act of 1980 authorized Idaho, Montana, Oregon and Washington to form the Northwest Power Planning Council (NWPPC), and directed the Council to coordinate the protection, mitigation and enhancement of fish and wildlife with the management and maintenance of the hydropower. Within two years, the fish and wildlife programme had formed. While initial efforts focused on improving fish survival at hydropower dams, the listing of salmon and steelhead as endangered species under the ESA in the early 1990s added urgency to habitat recovery (Loehman and Charney, 2011).

At the same time as the fish and wildlife programme underwent its fifth revision in 2000, the National Marine Fisheries Service published its guidance on salmon recovery efforts under the ESA. These two documents laid the groundwork for a coordinated programme of land and water acquisitions. In cooperation with the Bonneville Power Administration, the National Fish and Wildlife Foundation was selected to administer the Columbia Basin Water Transactions Program (CBWTP), which was established in 2002 to coordinate environmental allocation projects across a network of local implementing bodies known as 'qualified local entities' (CBWTP, 2017). The CBWTP therefore served as the umbrella organization to coordinate financing and capacity-building, building on prior experimentation with the 'water trust' model of instream flow restoration, statewide programmes and river conservancies. The CBWTP focuses on tributaries that are critical to the survival of anadromous and resident fish species, using stream discharge data to establish flow targets on specified stream reaches. The programme has selected a number of priority sub-basins with significant unmet demand, including: the Salmon in Idaho; the Bitterroot, Blackfoot, Clark Fork and Flathead in Montana; the Deschutes, Grande Ronde, John Day, Umatilla and Willamette in Oregon; and the Upper Columbia, Walla Walla and Yakima in Washington.

13.3   METHODOLOGY

13.3.1   WMRA Application

The Columbia Basin's extensive experience with water market implementation illustrates the completion of all steps in the WRMA framework, albeit to varying degrees of completeness and capacity. A range of market-based tools have been used to create open-market transactions since the late 1980s, with subsequent policy and institutional reforms improving conditions to be increasingly suitable for trading. The Columbia Basin is most accurately characterized as being in Stage 3 – monitoring and continuous review and assessment – while filling in the gaps and strengthening capacity for elements of Stages 1 and 2.

While transactions have been occurring for three decades, the CBWTP has been operating since 2003. In its lifetime, the programme has experimented with strategies to limit and reduce transaction costs, adapt to new information and address third-party effects, including socio-economic impacts on upstream communities. After an initial increase, the programme's expenditures have remained relatively stable over time (Figure 13.2). The process of review and assessment began early in the programme, with an extensive evaluation by Hardner and Gullison (2007) contributing to a better understanding of key factors that limited the ability of qualified local entities to complete water transactions.

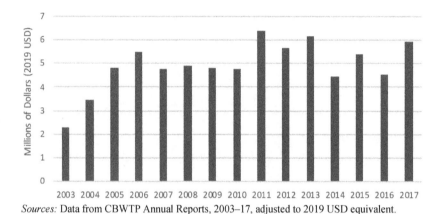

Sources: Data from CBWTP Annual Reports, 2003–17, adjusted to 2019 USD equivalent.

Figure 13.2   *Total annual expenditures by the CBWTP according to its annual reports, including money spent for water and implementation*

A follow-up survey in 2017 included questions related to hydrology, the market, capacity, monitoring and water rights. These categories align thematically with the prior 2007 survey carried out by Hardner and Gullison, and with the WMRA framework's fundamental market enablers (Figure 13.3). This conceptual alignment highlights the relevance of our survey data for understanding market readiness as defined by the WMRA framework. Discrepancies between the frameworks largely stem from the nature of the CBWTP as a non-profit, which operates as the buyer for environmental water. The presence of non-profits in the governance regime in the US differs from other locations, such as Australia's federal government-centred programmes (Garrick et al., 2009; Wheeler et al., 2013). These differences are further explored in the discussion section.

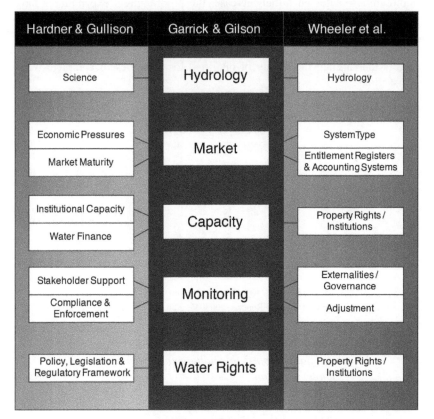

Figure 13.3   Comparison of categories of questions for the CBWTP surveys to the WMRA framework

13.3.2   Survey Details

We apply the WRMA framework to the results of our 2017 survey, which examined the drivers and barriers to water transactions in the CBWTP. The survey was administered to the CBWTP's network of qualified local entities (known as QLEs). Participants ranked barriers on a scale between 1 and 5, with 1 as 'not limiting', 3 as a 'manageable problem', and 5 as a 'major barrier'. Participants were asked to give an example of how each barrier has impeded their work. The survey contained 40 questions in total, including questions in which QLEs self-identified up to three responses for: (1) the strongest barriers they face today; (2) strong barriers that they confronted when first starting their position which they no longer face today; and (3) factors that are likely to emerge as barriers to flow transactions in the future.

Summary statistics of participants are found in Figure 13.4. Of the 30 individuals who completed the survey: four were agency staff, two were external consultants, and 24 were non-profit staff (including 20 project managers and 2 executive directors). Broken down by state: three participants were from Idaho, six from Montana, 10 from Washington, 11 from Oregon. The average experience of the participants working on flow transactions was 5.8 years. Twenty-seven of the individuals indicated that they are involved in other types of conservation projects (for example, riparian habitat restoration) in addition to water transactions.

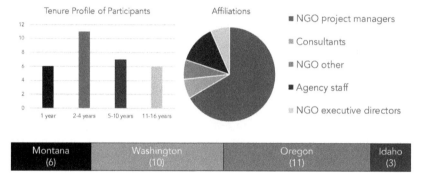

Figure 13.4   Summary of survey participants

13.4   RESULTS

13.4.1   Water Marketing Activity in the Columbia Basin

According to a water market database compiled by Donohew and Libecap (2010), which includes all types of water transactions between 1987 and 2009 in the western US, a total of 392 transactions occurred in the four Columbia Basin states. Transfers to the environment made up 47 per cent of total transactions. The next most frequent transfer was from agriculture to agriculture (30 per cent), followed by agriculture to urban (13 per cent).

According to CBWTP annual reports, the programme had a total of 232 active transactions and 540 total transactions as of 2017. On average, the CBWTP reports 36 new transactions per year, ranging from 30 to 49 annually (Figure 13.5). As of 2017, just under 100 000 acre-feet of water were protected in long-term deals, with another 80 000 acre-feet protected in temporary transactions. Projects in 2017 alone benefited nearly 500 miles of flows in 28 streams across the basin.

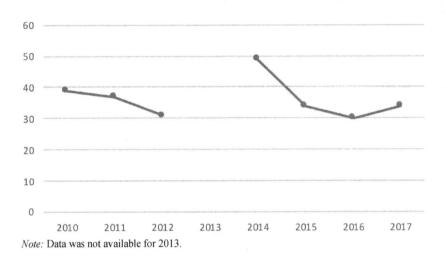

Note: Data was not available for 2013.

Figure 13.5   New transactions per year reported in CBWTP annual reports

13.4.2   Barriers to Transactions

Our survey addressed barriers in a number of different categories. First, we share the ranking of barriers relevant to the WRMA framework (Figure 13.3). Then, we summarize barriers identified by the respondents based on past experience and future expectations. Respondents were asked to identify the major barriers today, the major barriers encountered when they started their position (and whether they have persisted), and the barriers they anticipate in the next 3–5 years.

The presence of key factors associated with market readiness have not remained stable over time in the Columbia Basin. Instream flow transactions are not getting easier for the qualified local entities (QLEs), as more barriers were identified as impediments in 2017 than in 2007, and those identified in 2007 remain impediments today. The barriers identified in 2007 relate to the market and capacity, including uncoordinated and inadequate funding, the existence of adequate sellers, and high transaction costs. The seven additional impediments identified in our 2017 survey related to monitoring, capacity, quantification and administration of water rights. QLEs struggle to secure adequate funding, which exacerbates existing challenges related to monitoring capacity and meaningful evaluation. Problems related to water rights include inadequate enforcement, and administrative processing times that are too long. Socio-cultural barriers (categorized within 'monitoring') included a lack of community support, such as irrigators and irrigation districts that are concerned about their viability.

When QLEs were asked to self-identify barriers, three clear themes emerged related to socio-cultural factors, funding, and monitoring. These apply to the 'externalities/governance' and 'adjustment' categories of factors associated with market readiness in the WRMA framework. First, socio-cultural factors were the most frequently identified, and are a barrier to the majority of respondents. Examples include finding willing sellers, engaging with landowners though building and maintaining relationships, managing political and cultural resistance, and dealing with negative effects of transactions. Second, participants struggle to find reliable and sufficient funds to cover transactions, organizational capacity and monitoring. Third, monitoring and enforcement are increasingly burdensome, especially as QLEs accumulate a portfolio of transactions. A lack of monitoring data poses additional challenges for prioritizing new transactions. This relates both to the WRMA framework's 'accounting' market assessors, as well as the ability of the CBWTP to successfully make progress on Step 3: monitoring and continuous review and assessment.

Progress in overcoming barriers has been uneven, revealing that different entities are moving through Steps 2 and 3 of the WRMA framework with varying degrees of success. The set of barriers that have been overcome by QLEs were heterogeneous, with no obvious universal trends. The most common answer, from just over a quarter of participants, was that the personal knowledge and capacity of the respondent had improved, highlighting the importance of human capital associated with longer staff tenure and experience. The next most common answers were improved administrative processing times, and the development of beneficial collaborative partnerships, which were mentioned equally by participants. However, the same percentage of QLEs responded that they had seen no changes in barriers.

In contrast to the wide variety of barriers self-identified by QLEs, the majority agreed on two key barriers that were likely to emerge in the next 3–5 years: inadequate funds, and harder transactions. Funding is a concern to many participants, who are uncertain as to whether they will receive sufficient funds at the scale and pace necessary. QLEs are anticipating that transactions will become harder in the coming years due to a number of variables, including climate change, increased competition, fewer willing sellers, and deals that are more expensive and more complicated. These challenges are exacerbated by emerging trends such as an increase in the cost of water due to competition with growing cities and towns, and a lack of incentives for landowners at the prevailing prices. Some participants noted how the combination of uncertainties related to a changing climate and uncertainties related to funding poses additional challenges for strategic planning. This underscores the importance of Step 3 of the WRMA framework and emphasizes that review and assessment must be continuously ongoing.

13.5   DISCUSSION

In spite of achieving market readiness, and years of experience with transactions, our results demonstrate that non-profit actors in the Columbia Basin still face a number of barriers today. Interview responses show that: (1) culture matters; (2) institutional reform and implementation go hand in hand to build capacity at a local level; and (3) 'readiness' may never be complete.

13.5.1   Culture Matters

Cultural factors play a significant role in the success of transactions and should be seriously considered when designing water market arrangements. While the WMRA includes 'political acceptability of trade' under the 'adjustment' market assessor, certain geographies may need to expand the scope of their assessment beyond externalities to adequately and proactively prepare for socio-cultural barriers, such as push-back from farmers who perceive transactions as threatening to their cultural heritage. Building relationships is essential to successful engagement, including moving from short-term to long-term transactions and addressing fears of negative effects on landowners.

The Deschutes sub-basin of the Columbia is a noteworthy example of the importance of investing early in the multi-stakeholder processes, as high costs in the early stages of environmental water market development brought about high returns. The Deschutes River Conservancy invested in: (1) institutional water banking mechanisms and partnerships; (2) relationships with local irrigation districts; and (3) communication, policy forums and basin-level planning to garner public support (Garrick, 2015; Lieberherr, 2011). This investment paid off in high performance, as the Deschutes consistently sees high water recovery levels and relatively stable transaction costs (Garrick and Aylward, 2012).

13.5.2   The Need for Capacity at All Levels

While some barriers are universally challenging, most are locally specific. However, local institutional capacity is uneven both within and across state boundaries. This causes increased variability and volatility related to performance measures of water recovery and transaction costs (Garrick and Aylward, 2012). For example, the Salmon in Idaho saw improved performance over time due to investment in state- and watershed-level policy reforms (Garrick and Aylward, 2012). The gap is growing between strong and weak sub-basins, as those that lack institutional capacity are unable to overcome local barriers and improve their overall performance. Enabling conditions must be pursued and coordinated at basin, state and local levels.

Survey responses illustrate how barriers vary within states as well as across them. For example, when asked to rank whether irrigators or irrigation districts are concerned about their viability (including capacity for operations and maintenance), all respondents from Idaho identified this as a problem, while answers from both Oregon and Washington included the full range of possible responses from 1 to 5. As another example, several of the barriers that participants had overcome since starting their position were self-identified as significant barriers by other participants, revealing uneven progress within the basin. These barriers included challenges with agency processing times, public awareness and support, defining flow targets, project prioritization, insufficient staff, and inadequate monitoring and enforcement. Taken together, these examples reveal the need for state-wide capacity and coordination. In the Columbia Basin, financing from federal sources can play a key role in enabling state progress.

13.5.3   Readiness May Never be Complete

Though the Columbia River Basin is 'ready enough' for water trading, three decades of experience suggest that it may never be fully ready. The third step in the WRMA framework – monitoring and continuous review and assessment – must be met with sustained investment. Enabling conditions are necessary, but insufficient for transactions to scale. Key assessors of market readiness should be revisited, reassessed and improved as conditions change and the market matures.

In the Columbia Basin, the nature of transactions continues to evolve (Garrick et al., 2013), and successful transactions have created an additional capacity burden related to monitoring and enforcement (Garrick et al., 2009). After the low-hanging fruit has been picked, concerted effort is needed to overcome cultural barriers and prioritize transactions amidst changing environmental conditions. While readiness implies the presence of stable enabling conditions, experience suggests that capacity for adaptive governance is also needed for continued success.

13.6   CONCLUSION

As one of the world's oldest experiments in environmental water markets, lessons learned in the Columbia River Basin provide valuable insights. In this chapter, we reviewed the background, policy evolution and experience with markets of the Columbia Basin as they relate to Steps 1, 2 and 3 of the WMRA framework. We then introduced our 2017 survey on barriers to transactions. Cultural factors emerged as a significant barrier to transactions for the majority of participants, underscoring the need to consider cultural context in the early stages of water market development. Though some of the barriers identified in our survey were experienced broadly by participants, most barriers are locally specific, highlighting the importance of local institutional capacity. The results indicate that progress in overcoming barriers is uneven.

While the Columbia Basin has been ready for water transactions for decades, our results challenge the underlying concept of 'market readiness' by demonstrating how barriers related to the key fundamental market assessors have changed over time. As environmental water trading has evolved, the nature of the transactions has changed. Further progress is needed to address uncertainty related to climate change, insufficient finance, and inadequate monitoring and enforcement. Progress related to the market assessors in the WRMA framework may be non-linear and involve cyclical processes to revisit and strengthen basic enabling conditions as transactions move to scale.

REFERENCES

Anderson, T.L., and Leal, D.R. (1991). *Free Market Environmentalism.* Westview Press.

Aylward, B. (2008). *Water Markets: A Mechanism for Mainstreaming Ecosystem Services into Water Management?* IUCN Briefing Paper, Water and Nature Initiative. IUCN.

CBWTP (2003–17). *Columbia Basin Water Transactions Program Annual Reports.*

Charney, S. (2005). *Decades Down the Road: An Analysis of Instream Flow Programs in Colorado and the Western United States.* Colorado Water Conservation Board.

Colby, B.G. (1990). Transactions costs and efficiency in Western water allocation. *American Journal of Agricultural Economics, 72*(5), 1184–1192.

Cosens, B., McKinney, M., Paisley, R., and Wolf, A.T. (2018). Reconciliation of development and ecosystems: the ecology of governance in the International Columbia River Basin. *Regional Environmental Change, 18*(6), 1679–1692.

Donohew, Z., and Libecap, G. (2010). *Water Transfer Database.* University of California, Santa Barbara, CA.

Garrick, D.E. (2015). *Water Allocation in Rivers under Pressure: Water Trading, Transaction Costs and Transboundary Governance in the Western US and Australia.* Edward Elgar Publishing.

Garrick, D., and Aylward, B. (2012). Transaction costs and institutional performance in market-based environmental water allocation. *Land Economics, 88*(3), 536–560.

Garrick, D., Lane-Miller, C., and Mccoy, A.L. (2011). Institutional innovations to govern environmental water in the western United States: lessons for Australia's Murray–Darling Basin. *Economic Papers, 30*(2), 167–184.

Garrick, D., Siebentritt, M.A., Aylward, B., Bauer, C.J., and Purkey, A. (2009). Water markets and freshwater ecosystem services: policy reform and implementation in the Columbia and Murray–Darling Basins. *Ecological Economics*, *69*(2), 366–379.

Garrick, D., Whitten, S.M., and Coggan, A. (2013). Understanding the evolution and performance of water markets and allocation policy: a transaction costs analysis framework. *Ecological Economics*, *88*, 195–205.

Hand, B.K., Flint, C.G., Frissell, C.A., Muhlfeld, C.C., Devlin, S.P., et al. (2018). A social–ecological perspective for riverscape management in the Columbia River Basin. *Frontiers in Ecology and the Environment*, *16*(S1), S23–S33.

Hardner, J., and Gullison, R.E. (2007). *Independent External Evaluation of the Columbia Basin Water Transactions Program (2003–2006)*. Hardner and Gullison Consulting.

Hurley, P.T. (2013). Whose sense of place? A political ecology of amenity development. In W.P. Stewart, D.R. Williams and L. Kruger (eds), *Place-Based Conservation* (pp. 165–180). Springer.

Kirschbaum, R.L., and Lettenmaier, D.P. (1997). *Evaluation of the Effects of Anthropogenic Activity on Streamflow in the Columbia River Basin*. Water Resources Series, Technical Report, University of Washington, Seattle, WA.

Lieberherr, E. (2011). Acceptability of the Deschutes groundwater mitigation program. *Ecology Law Currents*, *38*, 25.

Loehman, E.T., and Charney, S. (2011). Further down the road to sustainable environmental flows: funding, management activities and governance for six western US states. *Water International*, *36*(7), 873–893.

Matheussen, B., Kirschbaum, R.L., Goodman, I.A., O'Donnell, G.M., and Lettenmaier, D.P. (2000). Effects of land cover change on streamflow in the interior Columbia River Basin (USA and Canada). *Hydrological Processes*, *14*(5), 867–885.

McCoy, A.L., Holmes, S.R., and Boisjolie, B.A. (2018). Flow restoration in the Columbia River Basin: an evaluation of a flow restoration accounting framework. *Environmental Management*, *61*(3), 506–519.

Neuman, J.C., and Chapman, C. (1999). Wading into the water market: the first five years of the Oregon Water Trust. *Journal of Environmental Law and Litigation*, *14*, 135.

Neuman, J., Squier, A., and Achterman, G. (2006). Sometimes a great notion: Oregon's instream flow experiments. *Environmental Law*, *36*, 1125.

Northwest Power and Conservation Council (NPCC) (2014). *Columbia River Basin Fish and Wildlife Program 2014*.

Settre, C., and Wheeler, S.A. (2017). Rebalancing the system: acquiring water and trade. In A. Horne, A. Webb, M. Stewardson, B. Richter and M. Acreman (eds), *Water for the Environment: From Policy and Science to Implementation and Management* (pp. 399–419). Elsevier.

Thompson, B. (1993). Institutional perspectives on water policy and markets. *California Law Review*, *81*, 671.

Wheeler, S.A., Garrick, D., Loch, A., and Bjornlund, H. (2013). Evaluating water market products to acquire water for the environment in Australia. *Land Use Policy*, *30*(1), 427–436.

Wheeler, S.A., Loch, A., Crase, L., Young, M., and Grafton, R.Q. (2017). Developing a water market readiness assessment framework. *Journal of Hydrology*, *552*, 807–820.

14. Canterbury, New Zealand case study of the water market readiness framework assessment

Julia Talbot-Jones and R. Quentin Grafton

14.1 INTRODUCTION

In 1986 New Zealand made global headlines for being the first country in the world to comprehensively introduce an environmental market to address the over-use of fisheries (Lock and Leslie, 2007). The comprehensive Quota Management System identified a total allowable catch and 'grandparented' individual transferable quotas (ITQs), initially in terms of absolute tonnages of fish, to fishers based on their individual catch histories. ITQ holders were then able to trade and exchange their rights depending on the value each quota holder placed on their respective tonnage (Rees, 2005).

While not without its challenges, especially in terms of the initial quota allocations, the introduction of ITQs for managing New Zealand's fisheries has been a positive development in terms of biological and economic sustainability (Costello et al., 2008; Newell et al., 2005, 2007). Given this, it would be reasonable to expect that the environmental market model would have been adopted for other natural resources that are struggling with over-use, such as water (Leonard et al., 2019). Yet, this has not been the case. Instead, even in semi-arid regions, such as Canterbury on the east cost of the South Island, traditional supply-side approaches to water scarcity – diverting surface water, constructing dams and reservoirs, and pumping groundwater – have remained the most widely utilised response to meeting increases in demand (Jenkins, 2013, 2018).

At least in part, this approach to management has been driven by the fact that, until relatively recently, augmenting supply was viewed as an adequate approach to meeting water demands without imposing negative costs on the water system as a whole. From the 1990s onwards, however, land use in Canterbury has been changing in response to increased dairy demand, and today the region includes 70 per cent of New Zealand's irrigated land and accounts for approximately 60 per cent of all water allocated for consumptive use in New Zealand (Painter, 2018). This, alongside population and economic growth in urban areas, has placed increasing stress on the region's water systems (Jenkins, 2015).

Like other semi-arid regions in New Zealand, Canterbury is now at a cross-roads (Jenkins, 2018). It cannot resolve its emerging water issues using supply-side approaches alone. Instead, it must complement the traditional mechanisms with demand-side management tools that promote the efficient use of existing supply. These tools could include such approaches as providing incentives for urban households to use less water, finding ways to reuse municipal effluent, or by creating institutions that would make water allocations more flexible in rural areas. As rural users are the greatest consumptive users of water in Canterbury, establishing a system that manages their growing demands could deliver benefits to all water users.

This chapter considers how a water market could work in the Canterbury context by applying the water market readiness assessment (WMRA) framework to the region (Wheeler et al., 2017). We assess how 'ready' Canterbury is to use economic market mechanisms to respond to increasing water demands, and assesses the broader applicability and value of the WRMA framework across different contexts.

14.2 THE CASE STUDY: CANTERBURY, NEW ZEALAND

The Canterbury region is located on the central east coast of the South Island. The region stretches west across plains from the coast to the foothills of the Southern Alps, up to Marlborough, and south to Otago, covering an area of 44 915 km^2. The region is defined by its braided rivers, high country, coastal lakes, lowland streams, and wetlands. It also has the greatest potential evapotranspiration deficit of any New Zealand region. For this reason, how water is allocated across and between competing uses is of importance for both economic and environmental reasons (Saunders and Saunders, 2012).

In recent years, demand for surface water and groundwater has increased significantly in the region as more land has been converted to irrigated agriculture (Saunders and Saunders, 2012). In 1982, the region's irrigated area was estimated at about 100 000 ha (Dommisse, 2005); by 2015 this had increased fivefold to 507 000 ha (Brown, 2016). In addition, Canterbury generates 24 per cent of New Zealand's power through hydroelectricity and has 65 per cent of the country's hydro-storage (Jenkins, 2018). The region's world-famous braided rivers also provide a recreational playground for users, and valuable habitat for endangered and threatened wildlife. In addition, the rivers and aquifers provide a high-quality water supply to Canterbury's major city, Christchurch.

These competing uses have put pressure on the region's water systems and made clear that the region's hydrology is not well understood outside of the Christchurch aquifer system (Dench, 2017; Duncan, 2016). Most river and aquifer systems are now at or near their sustainability limits, with ten of Canterbury's 36 groundwater zones exceeding their allocation limits and three exceeding 80 per cent of the allocation limit (Jenkins, 2018). These high levels of groundwater extraction, coupled with nutrient loading, have also affected water quality, with direct impacts on aquatic ecosystem health and human health (Environment Canterbury, 2018; Painter, 2018).

In 1998, a reform process began in Canterbury to respond to growing water scarcity and concerns over water quality. The resulting Canterbury Water Management Strategy aimed '[t]o enable present and future generations to gain the greatest social, economic, recreational and cultural benefits from our water resources within an environmentally sustainable framework' (Environment Canterbury, 2010, p. 6). Although elements of these reforms have the potential to improve outcomes for the Canterbury region, little attention has been given to considering price or market-based mechanisms as a way of addressing declining water availability and water quality (Jenkins, 2013; Kirk et al., 2017). Thus, there remains an opportunity to understand how further reform would aid in the delivery of efficient and cost-effective governance solutions.

14.2.1   Step One: Background Context

Most water systems in New Zealand are managed using a top-down system of decision-making as prescribed by the Resource Management Act (RMA) 1991, New Zealand's principal legislative framework for managing natural and physical resources. Under the RMA, the central government provides overall direction by outlining National Policy Statements and setting National Environmental Standards, but devolves decision-making to the lowest levels of government at which matters can be appropriately considered and adminis-tered. In Canterbury, this means that control and management of water is con-centrated in the hands of a regional council, Environment Canterbury (ECan).

ECan's main role is to promote the sustainable management of natural and physical resources, primarily under the RMA and the Local Government Act 2002. The RMA grants ECan the authority to award water users permits to take or use water; activities that would otherwise contravene section 14 of the RMA. These permits are essentially a form of licensing under command-and-control regulation, and are known as resource consents. Resource consents can be granted for a period of up to 35 years; however, if no period is specified in the consent, it is deemed to be granted for five years from its date of commencement. The resource consent may lapse or be cancelled by ECan if the activity covered by the consent is not undertaken within a five-year period.

Resource consents cannot be traded easily under legislation, for several reasons. First, although consents for water are not connected to land, they can only be held by individuals or groups who own or occupy land adjacent to the water source, or who have access to pre-existing infrastructure for water transfer. This increases transaction costs and constrains the type of trading that can take place. Second, although section 136 of the RMA states that consents can be transferred from one owner-occupier to another by applying to the responsible authority, the legislation states that the consents cannot be traded at will as the property of an individual. This again increases transaction costs, by creating ambiguity around how the property rights are defined, defended and able to be divested.

Despite these barriers, low levels of transfer do occur in Canterbury, suggesting that there is some value in water trades between consent holders. Between 2013 and 2016, for example, 27 applications for full transfer were placed with ECan (17 granted), 23 applications for partial transfer and limited duration were submitted (21 granted), and 131 for partial transfer and full duration were applied for (104 granted) (Sharpe, 2016).

ECan processes applications for transfers and new applications for consent on a first come, first served basis. This means that ECan processes applications in the order they are received and rejects new applications once a system is fully allocated, irrespective of whether or not they are likely to deliver better economic and environmental outcomes than those already granted (Aoraki Water Trust v Meridian Energy Limited [2005] 2 NZLR 268). This approach, coupled with poor information and data management, has led to the over-allocation of available supply and declining water quality across the region (Jenkins, 2018; Snelder et al., 2018; Wright, 2013).

For many water users in Canterbury, declining water quality and quantity is of growing concern. The possibility that the irrigated area may increase does little to assuage their fears, and conflict has arisen between farmers (who are frequently painted as the perpetrators), urban residents, environmentalists and some members of Ngāi Tahu (local Māori iwi). The response of the central government has been to appoint independent councillors to oversee allocation decisions, which has been viewed as a breach of democratic process (Brower, 2010).

At the local government level, ECan has moved to devise and implement a series of rules and regulations focused on reducing levels of allocation. However, the pace of change is slow and asymmetrical; most of the legislative effort has gone into increasing supply through diverting more water from rivers into irrigation schemes or constructing more dams and reservoirs (Jenkins, 2013, 2015). In sum, for Canterbury to remain resilient in the face of substantial disruptions of water supplies, there could be some benefit in adopting a wider suite of institutional measures so that cities, farms, industries and environmental interests can be sustained within the existing water supply.

14.2.2 Step Two: Evaluation, Development and Implementation

Water markets represent a potentially useful tool for reallocation. Indeed, given that Canterbury already has an existing consent system, has allowed consent transfer, and implemented monitoring technology and minimum flows, this suggests that Canterbury is already at Step Two of the WMRA framework. However, several institutional barriers that increase transaction costs are limiting the development of an efficient water trading system. For instance, the classification of consents as rights in the RMA is problematic. Section 122 of the RMA declares consents to be neither real nor personal property (section 122(1)); however, this is immediately complicated by the following five sub-sections of section 122, which bring property law to bear 'as if' a consent is personal property (Barton, 2010). Thus, because water markets are more efficient when property rights are well defined, the current classification of consents under the RMA fails to provide a strong foundation for establishing rights and their corresponding duties and responsibilities.

Instead, the current consent structure under the RMA implies that consents confer a partial right to access and withdrawal of water (Schlager and Ostrom, 1992), but not one of ownership or full control (Palmer, 2015). This ambiguity can be attributed to the long-standing disagreements between the Crown and Māori over ownership and control of water that have been driven by inconsistent translations of Te Tiriti o Waitangi (Treaty of Waitangi), one of New Zealand's founding documents (Durie, 1993; Memon and Kirk, 2012; Tau, 2017; Waitangi Tribunal, 2012). The issues around ownership and control remain unresolved, and continue to act as a barrier to the establishment of alternative institutional arrangements such as pricing or market-based mechanisms, that require well-defined property rights (Murray et al., 2014; Tau, 2017).

Rules within the regional plan also create barriers to trade. For example, ECan's Land and Water Regional Plan requires all consent transfers to be approved by ECan; a process which is costly to consent holders wishing to undertake a transfer (Environment Canterbury, 2016). Likewise, section 4.71(d) of ECan's Land and Water Regional Plan requires all consent holders in over-allocated surface water catchment or groundwater zones to surrender a proportion of their consent when participating in trade. This creates a disincentive for consent holders to transfer their consents unless the benefits of doing so clearly outweigh the substantive financial and time-consuming costs. The fact that the number of applications for transfers has declined since this rule was implemented in 2016 supports a hypothesis that the imposition of the rule has increased the costs of participating in trade and exchange (see www .hydrotrader.co.nz).

Poor monitoring, enforcement and data analysis also inhibit a smooth shift from the status quo. Although all consent holders were required to install a water meter to measure and report water takes of 10 litres/sec or more in 2010, and water takes of 5 litres/sec or more needed to be measured and reported from late 2016 (Resource Management (Measurement and Reporting of Water Takes) Regulations, 2010)), there remains a high degree of uncertainty regarding water use patterns in the region (Dodson, 2015; Macdonald, 2018). Although ECan has collected data on use patterns relative to allocation from 2012, there is minimal monitoring and evaluation of consent holder behaviour (Dodson, 2015). As a result, estimates of water use, which would suggest that users are not using their full consents, are difficult to validate (see Glubb and Durney, 2014; Ministry for the Environment and Stats NZ, 2017).

It is only through the resolution of cultural concerns, the improvement of data analysis, and further institutional change, that a water market is likely to be formally implemented in Canterbury. Nevertheless, there is some evidence that there is demand for trade and that the trade and exchange of consents could deliver net benefits to users. For instance, Hydrotrader, a private company, acts as a broker for consent holders in Canterbury who want to trade consents. It matches prospective buyers and sellers who then apply to ECan for the official transfer of consents. Between 2007 and 2016, Hydrotrader steadily increased the number of transfers it brokered between users; however, the 2016 regional plan rule changes that required consent holders to have their water tested and to relinquish some of their allocation when applying for consent transfer correlates with a fall in applications for trade and exchange. This indicates that ECan's rule changes have significantly affected transaction costs for consent transfer (see www.hydrotrader.co.nz).

In cases where trade can be facilitated at low cost, there is evidence that consent holders in Canterbury do choose to engage in trade and exchange. Under the RMA, the definition of water excludes water contained in a pipe. This means that transfers under irrigation schemes are not regulated by the RMA, and members of irrigation schemes can make their own rules or arrangements relating to transferability of entitlements (Milner-White, 2010). In cases where this has been possible, there is clear evidence that entitlement holders have a preference for trade and that significant capital gains can be made through trading (Rockpoint, 2012). In south Canterbury, for example, the collectively owned Opuha Dam allows some trade and exchange of water shares between investors under certain conditions. Opened in 2000, the original investors of the Opuha Dam (farmers, the council and independent shareholders) were required to provide an initial deposit of $50, followed by a call of $200 for each share (one share irrigates 4 hectares, $62.50/ha). By 2006 the scheme was fully subscribed, and water shares traded at prices exceeding $4000 ($1000/ha) (Rockpoint, 2012).

Demand for market-based mechanisms is also evident in the water quality space. Although operating outside the Canterbury region, New Zealand is one of the few places in the world that has an operative non-point source to non-point source water quality trading market (Shortle, 2012). The Lake Taupō nitrogen trading scheme, which became fully operative in July 2011, consists of three key market components: a cap, a public fund for buybacks, and trading (Duhon et al., 2015). Nitrogen allowances were initially grandparented to participants, and permit holders have since been required to comply with a management plan that regulates and constrains permit holders' behaviour. Permit holders are allowed to trade allowances with one another or with an independent trust, the Lake Taupō Protection Trust, which is charged with permanently reducing nitrogen losses by 20 per cent over time. The Trust can achieve its objective through the purchase and conversion of land, or through the purchase and permanent retirement of farmers' nitrogen allowances, using a public fund with contributions from local, regional and national communities. To make a trade, both the buyer and seller must submit an updated nitrogen management plan for Council approval. Under the market-based model, the reduced nitrogen target has been achieved ahead of schedule, making the management scheme a success; however, it should be noted that 90 per cent of the reduction has been achieved through purchases by the Trust rather than farmers trading discharge allowances (Jenkins, 2018).

A second water quality market on Lake Rotorua provides an interesting counter-case because the trading system has failed to develop, despite a weakly monitored freeze on nitrogen leaching from farms since 2005. Using the criteria outlined in the WMRA framework, reasons for the delayed implementation could be attributed to the facts that Lake Rotorua has more severe water quality problems than Taupō, and has a more complex groundwater hydrology. The lake is one of 16 in the area, and it has been estimated that 53 per cent of the nitrogen reaches the lake via groundwater with lags of up to 120 years (Anastasiadis et al., 2011). From an institutional perspective, the framework proposed for Lake Rotorua attempts to learn from the Lake Taupō experience and refine the nutrient-trading framework, but in doing so this introduces further complexities (Kerr et al., 2012). Plans for the Rotorua framework, for instance, seek to reduce transaction costs by using a self-reporting system that avoids the need for approved farm management plans or regional council approval for trades. However, this approach requires the imposition of more certain and swift non-compliance penalties and more monitoring and enforcement by local authorities; an institutional element that has its own associated costs and complexities. Trying to balance the levels of transaction costs across the various elements of the institutional framework are likely to remain a core challenge. The lessons that come out of these two water quality trading examples provide insights into the challenges that might arise in the Canterbury region. For a water quantity market to work in Canterbury, the benefits of trading need to be clearly relayed to the public, to iwi, to consent holders and to policy analysts. Currently, Canterbury has unresolved property rights issues concerning ownership of water that are inhibiting both a price being placed on water and the consideration of alternative arrangements (Tau, 2017). To enable the adoption of economic instruments, property rights need to be clearly defined, and transaction costs and barriers to trade need to be reduced.

14.2.3 Step Three: Monitoring and Review

The application of the WMRA framework to Canterbury indicates that the region has reached Step Two of the framework, but the barriers to reaching Step Three are substantial. Nevertheless, it is useful to consider how a water market might operate in terms of monitoring and review if it were to be implemented.

The existing metering system provides ECan with the ability to collect data and monitor behaviour. At present not all consents are regularly monitored; nevertheless, greater resourcing could address this compliance gap. Whether greater resourcing could address the weakness of New Zealand's local government to enforce adverse behaviour is unclear (Brown et al., 2016; Ministry for the Environment, 2016). Local governments face several challenges regarding enforcement, which converge on four main themes: antiquated legislation, poor resourcing, a lack of regulatory independence, and limited auditing and oversight (Brown, 2017). This suggests that to spearhead a comprehensive monitoring and review process necessary for an effective trading system, appropriate effort would have to be assigned to compliance and monitoring across all four themes, as well as to the promulgation of law and policy (Horne and Grafton, 2019).

We contend that flexibility and adaptability would need to be built into the market design to allow for adaptation and adjustment of the trading and use rules over time. As shown in the case of New Zealand fisheries, the initial Quota Management System required adjustment three years after implementation, particularly in terms of the size of the cap, and how harvesting rights were initially allocated, and to whom (Rees, 2005). Initially, ITQs were allocated in terms of tonnes of harvested fish that subsequently was revised to a percentage of the total allowable catch (TAC). This required a substantial level of compensation in one fishery where the TAC had been set above the sustainable level. The subsequent quota buybacks were costly for government, but more than 30 years on they have resulted in a system that remains a model for other fishery management systems around the world.

14.3   DISCUSSION

Guided by the WRMA framework, our analysis suggests that Canterbury is currently sitting at Step Two of the water market readiness process. Using the criteria outlined by the framework we conclude that the framework allows for a comprehensive assessment of most institutional barriers in Canterbury, except for the social and cultural acceptability of trade. Importantly, these social and cultural elements need to be part of any broader water reform framework, as they often shape and determine the type of institutional pathways that can evolve and be effectively implemented (Grafton et al., 2019). For regions that have water systems of cultural importance, such as Canterbury, or have stakeholders with concerns about assigning a market price to water, not including these categories in the WRMA framework could result in some analysts under-estimating the importance of these market factors (Grafton et al., 2019; Nikolakis et al., 2013).

Nevertheless, should a water market eventually be established in Canterbury, the WMRA framework offers valuable guidance for design and implementation. For instance, the framework draws attention to the fact that Canterbury already has many of the core market requirements in place: extraction limits, consents and modes of consent transfer. The framework also highlights that the main barriers to trade are high transaction costs and unresolved cultural issues around ownership and property rights.

One important consideration for assessing the readiness of Canterbury for a water market that the framework does not encourage analysts to consider is the fact that in some cases a water market could result in increased water consumption; precisely the opposite goal of market design (Young, 2014). In Canterbury, current estimates of monitored water use are significantly below the allocation limits (see Glubb and Durney, 2014). If trade and exchange is permitted and the cap is set below the current allocation limit, but above current use levels, it is possible that trade will result in increased water use as partial or full consents begin to be transferred from low- to high-value users. The availability of 'new' water could also lead to unanticipated increases in irrigated area and an increased demand for water. Ultimately, it is possible that the savings anticipated from implementing efficiency measures fail to result in increased water availability, leaving the region worse off overall (Grafton et al., 2018).

For this reason, setting the cap at the appropriate level will be critical to ensure that an efficient and equitable water market can be established in Canterbury. Setting the cap too low is likely to put unnecessary pressure on users, but setting it too high could further encourage over-use. The good thing is that, rather than being a barrier to establishing a market, the challenge of setting the cap could be seen as an opportunity to help navigate some of the complex social and cultural issues currently preventing the adoption of market mechanisms. For instance, to uphold the partnership principles of the Treaty of Waitangi, the annual or seasonal cap could be decided jointly by appointees from local iwi and ECan, and individual allocations could then be grandparented based on existing consent allocations. This approach to co-governance would not be without precedent, and could draw on existing water governance models that are designed to facilitate shared decision-making between iwi and local authorities (Muru-Lanning, 2012; Salmond, 2014). Likewise, experiences from fisheries management could help to inform how cultural rights could be allocated, and how a system of trade could be designed that keeps transaction costs low (Bess, 2001; Costello et al., 2008). Combining these lessons from existing environmental market models with insights from the WRMA framework will help to ensure that the market criteria can be met efficiently within the local context.

14.4 CONCLUSION

In order for Canterbury to meet future water demand, introducing a range of complementary economic tools that increase the flexibility of water delivery could be useful. To date, ECan has focused on augmenting supply and regulating through limits, rather than using economic tools to shape behaviour (Jenkins, 2018). We contend that, in some cases, introducing economic tools such as water trading could facilitate the reallocation of water to meet the forecasted increases in demand. Given this, the water trading framework proposed here could play a useful role in determining how a market could be designed and implemented, and ultimately encourage the conservation and stewardship of water supplies.

Applying the WRMA framework to the Canterbury region focuses attention on the institutional elements needed for a water market to operate effectively and efficiently. We contend that the preliminary market structures that are currently in place mean that Canterbury is at Step Two of the water market readiness framework; between 'market initiating change I' and 'market initiating change II'. Key challenges for a water trading system in Canterbury are: the identification of a system that supports the transfer of consents; the setting of a widely accepted cap on water use (including estimates of water consumption); flexibility to revise water use and trading rules; proper consideration of the rights of local iwi within the water trading and water governance rules (Taylor et al., 2019); and the effective integration of water markets within overall water governance reform (Grafton et al., 2019).

REFERENCES

Anastasiadis, S., Nauleau, M.-L., Kerr, S., Cox, T., and Rutherford, K. (2011). *Does Complex Hydrology Require Complex Water Quality Policy?* NManager simulations for Lake Rotorua (No. 11–14). Motu Working Paper. www.motu.org.nz.

Barton, B. (2010). Property rights created under statute in common law legal systems. In A. McHarg, B. Barton, A. Bradbrook and L. Godden (eds), *Property and the Law in Energy and Natural Resources* (pp. 80–99). Oxford University Press.

Bess, R. (2001). New Zealand's indigenous people and their claims to fisheries resources. *Marine Policy, 25*(1), 23–32.

Brower, A. (2010). Environment Canterbury (Temporary Commissioners and Improved Water Management) Act 2010. *New Zealand Journal of Environmental Law, 14*, 309–321.

Brown, M.A. (2017). Last line of defence: a summary of an evaluation of environmental enforcement in New Zealand. *Policy Quarterly, 13*(2), 36–40.

Brown, M.A., Peart, R., and Wright, M. (2016) *Evaluating the Environmental Outcomes of the RMA*. Wellington, NZ, Environmental Defence Society.

Brown, P. (2016). *Canterbury Detailed Irrigated Area Mapping*. Aqualinc. www.api .ecan.govt.nz.

Costello, C., Gaines, S.D., and Lynham, J. (2008). Can catch shares prevent fisheries collapse? *Science, 321*(5896), 1678–1681.

Dench, W. (2017). Identifying changes in groundwater quantity and quality resulting from border-dyke to spray irrigation conversion. Masters dissertation, University of Canterbury.

Dodson, J. (2015). *Naturalising Flows and Estimating Water Abstraction in Canterbury*. Environment Canterbury. www.api.ecan.govt.nz.

Dommisse, J. (2005). *A Review of Surface Water Irrigation Schemes in Canterbury: Their Development, Changes with Time and Impacts on the Groundwater Resource*. Environment Canterbury. www.api.ecan.govt.nz.

Duhon, M., McDonald, H., Kerr, S. (2015). *Nitrogen Trading in Lake Taupo: An Analysis and Evaluation of an Innovative Water Management Policy*. Motu Working Paper. http://motu-www.motu.org.nz/wpapers/15_07.pdf.

Duncan, R. (2016). Ways of knowing – out-of-sync or incompatible? Framing water quality and farmers' encounters with science in the regulation of non-point source pollution in the Canterbury region of New Zealand. *Environmental Science and Policy, 55*, 151–157. https://doi.org/10.1016/J.ENVSCI.2015.10.004.

Durie, E.T. (1993). Will the settlers settle – cultural conciliation law. *Otago Law Review, 8*(4), 449–465.

Environment Canterbury (2010). *Canterbury Water Management Strategy*. https://ecan .govt.nz/your-region/plans-strategies-and-bylaws/canterbury-water-management -strategy/.

Environment Canterbury (2016). *Canterbury Land and Water Regional Plan*. https:// www.ecan.govt.nz/your-region/plans-strategies-and-bylaws/canterbury-land-and -water-regional-plan/.

Environment Canterbury (2018). *Annual Groundwater Quality Survey*. https://ecan .govt.nz/get-involved/news-and-events/2019/groundwater-quality-survey-released/.

Glubb, R., and Durney, P. (2014). *Canterbury Region Water Use Report for the 2013/14 Water Year*. Environment Canterbury. www.api.ecan.govt.nz.

Grafton, R.Q., Garrick, D., Manero, A., and Do, T.N. (2019). The water governance reform framework: overview and applications to Australia, Mexico, Tanzania, USA and Vietnam. *Water*, *11*(1), 137–159.

Grafton, R.Q., Williams, J., Perry, C.J., Molle, F., Ringler, C., Steduto, P., et al. (2018). The paradox of irrigation efficiency. *Science*, *361*(6404), 748–750. https://science .sciencemag.org/content/361/6404/748.

Horne, J., and Grafton, R.Q. (2019). The Australian water markets story: incremental transformation. In J. Luetjens, M. Mintrom and P. Hart (eds), *Successful Public Policy: Lessons from Australia and New Zealand* (pp. 165–190). ANU Press. https:// doi.org/10.22459/spp.2019.07.

Jenkins, B. (2013). *Progress of the Canterbury Water Management Strategy and Some Emerging Issues*. IDEAS Working Paper Series from RePEc. https://ir.canterbury.ac .nz/handle/10092/9084.

Jenkins, B. (2015). New Zealand water pricing. In A. Dinar, V. Pochat and J. Albiac-Murillo (eds), *Water Pricing Experiences and Innovations* (pp. 263–288). Springer International Publishing.

Jenkins, B. (2018). *Water Management in Canterbury*. Springer International Publishing.

Kerr, S., McDonald, H., and Rutherford, K. (2012). *Nutrient Trading in Lake Rotorua: A Policy Prototype*. Motu Working Paper. www.motu.org.nz.

Kirk, N., Brower, A., and Duncan, R. (2017). New public management and collaboration in Canterbury, New Zealand's freshwater management. *Land Use Policy*, *65*, 53–61.

Leonard, B., Costello, C., and Libecap, G.D. (2019). Expanding water markets in the western United States: barriers and lessons from other natural resource markets. *Review of Environmental Economics and Policy*, *13*(1), 43–61. https://doi.org/10 .1093/reep/rey014.

Lock, K., and Leslie, S. (2007). New Zealand's Quota Management System: a history of the first 20 years. *SSRN Electronic Journal*. https://papers.ssrn.com/sol3/papers .cfm?abstract_id=978115.

Macdonald, N. (2018). Critical shortage of good data on use of precious water resource. https://www.stuff.co.nz/environment/108707532/how-do-we-use-water-we-still -dont-really-know.

Memon, P.A., and Kirk, N. (2012). Role of indigenous Māori people in collaborative water governance in Aotearoa/New Zealand. *Journal of Environmental Planning and Management*, *55*(7), 941–959.

Milner-White, G.R. (2010). Water management reform – what role does the market have to play? *Water NZ Conference* (p. 20). https://www.waternz.org.nz.

Ministry for the Environment (2016). *Compliance, Monitoring and Enforcement by Local Authorities under the Resource Management Act 1991*. Wellington, NZ, Ministry for the Environment. http://www.mfe.govt.nz.

Ministry for the Environment and Stats NZ (2017). *New Zealand's Environmental Reporting Series: Our Fresh Water 2017*. Wellington, NZ, Ministry for the Environment. www.mfe.govt.nz and www.stats.govt.nz.

Murray, K., Sin, M., and Wyatt, S. (2014). *The Costs and Benefits of an Allocation of Freshwater to Iwi*. Sapere Research Group. www.iwichairs.maori.nz.

Muru-Lanning, M. (2012). The key actors of Waikato River co-governance. *AlterNative: An International Journal of Indigenous Peoples*, *8*(2), 128.

Newell, R., Papps, K., and Sanchirico, J.N. (2007). Asset pricing in created markets for fishing quota. *American Journal of Agricultural Economics*, *89*(2), 259–272.

Newell, R., Sanchirico, J.N., and Kerr, S. (2005). Fishing quota markets. *Journal of Environmental Economics and Management*, *49*(3), 437–462.

Nikolakis, W.D., Grafton, R.Q., and To, H. (2013). Indigenous values and water markets: survey insights from northern Australia. *Journal of Hydrology*, *500*, 12–20.

Painter, B. (2018). Protection of groundwater dependent ecosystems in Canterbury, New Zealand: the Targeted Stream Augmentation Project. *Sustainable Water Resources Management*, *4*(2), 291–300.

Palmer, S.G. (2015). Ruminations on the problems with the Resource Management Act. In *Annual Conference of New Zealand Planning Institute* (p. 19). https://www.planning.org.nz/Attachment?Action=Download&Attachment_id=3538.

Rees, E. (2005). In what sense a fisheries problem? Negotiating sustainable growth in New Zealand's fisheries. PhD dissertation, University of Auckland, New Zealand.

Rockpoint (2012). *Irrigation in New Zealand*. Welllington, NZ, Rockpoint. www.rockpoint.co.nz.

Salmond, A. (2014). Tears of Rangi: water, power, and people in New Zealand. *HAU: Journal of Ethnographic Theory*, *4*(3), 285–230.

Saunders, C., and Saunders, J. (2012). *The Economic Value of Potential Irrigation in Canterbury*. Lincoln, NZ, Agribusiness and Economics Research Unit. https://researcharchive.lincoln.ac.nz/handle/10182/6973.

Schlager, E., and Ostrom, E. (1992). Property rights regimes and natural resources: a conceptual analysis. *Land Economics*, *68*(3), 249–262.

Sharpe, S. (2016). *Water Trading in New Zealand*. Victoria University of Wellington Legal Research Paper. https://papers.ssrn.com/sol3/papers.cfm?abstract_id=2979730.

Shortle, J.S. (2012). *Water Quality Trading in Agriculture*. Paris, France, OECD. http://www.oecd.org/dataoecd/5/1/49849817.pdf.

Snelder, T.H., Larned, S.T., and McDowell, R.W. (2018). Anthropogenic increases of catchment nitrogen and phosphorus loads in New Zealand. *New Zealand Journal of Marine and Freshwater Research*, *52*(3), 336–361.

Tau, T.M. (2017). *Water Rights for Ngāi Tahu: A Discussion Paper*. Canterbury University Press.

Taylor, K., Longboat, S., and Grafton, R.Q. (2019). Whose rules? Principles of water governance, rights of Indigenous Peoples, and water justice. *Water*, *11*(4), 809.

Waitangi Tribunal (2012). *The Interim Report on the National Freshwater and Geothermal Resources Claim*. www.waitangitribunal.govt.nz.

Wheeler, S.A., Loch, A., Crase, L., Young, M., and Grafton, R.Q. (2017). Developing a water market readiness assessment framework. *Journal of Hydrology*, *552*, 807–820.

Wright, J. (2013). Water quality in New Zealand: land use and nutrient pollution. Wellington, NZ, Parliamentary Commissioner for the Environment. www.pce.parliament.nz.

Young, M.D. (2014). Designing water abstraction regimes for an ever-changing and ever-varying future. *Agricultural Water Management*, *145*, 32–38.

15. Lessons from water markets around the world

Sarah Ann Wheeler

15.1 INTRODUCTION

Many argue that there is a coming crisis in global water management (Barbier, 2019), and the use of property rights, and water markets in particular, will increasingly be given more attention and emphasis as water becomes scarcer in the face of future climate change, increasing populations, changing demand for water and increased environmental needs. The case studies in this book span six continents (Africa, Asia, Europe, South America, North America and Oceania), across 28 regions and 20 countries, and focus on how various countries are grappling with the issues of water scarcity, and trying to develop a range of strategies to deal with it. This chapter provides a summary of all the case studies, and some lessons learned.

15.2 WMRA CASE STUDY SUMMARY

Tables 15.1 and 15.2 below provide a summary of the application of the water market readiness assessment (WMRA) framework to 28 regions (in 20 countries) across six continents in the world. As a reminder of the three key institutional factors that are needed as a prerequisite for establishing water markets (Wheeler et al., 2017), they include three steps:

- Step 1: Enabling institutions. Defining the total resource pool available for consumptive use and hydrological factors of use; and evaluating the current institutional, legislative, planning and regulatory capacity to facilitate water trade.
- Step 2: Facilitating gains from trade. Developing clear and consistent trading rules; assessing benefits and costs of market-based reallocation, for example, numbers of individuals who can trade (versus adoption of trade); homogeneity of water use, adaptation benefits, cost of water reform, ongoing trade transaction costs, and assessment of externalities.

- Step 3: Monitoring and enforcement. Use of water markets and water extractions need ongoing monitoring and enforcement to ensure compliance, as well as continued development of trade enabling mechanisms, including: seeking to limit/reduce transaction costs, scanning for unanticipated externalities, developing new market products and then implementing, if needed, new legislative changes and planning requirements.

Appendix Table 2A.1 in Chapter 2 provided a list of various questions to highlight water market enabling and constraining factors.

15.2.1 Case Studies in Africa and Asia

Table 15.1 summarises the outcome of applying the WMRA framework to countries from Africa and Asia. It is shown that only one country – China – that was a case study in these continents has gone past Step One, and has established property rights and strong independent water institutions (a lack of strong governance impartiality is a key issue for many of these countries). Although all countries have implemented some form of water legislation to address scarcity issues, very few Asian countries have unbundled water from land rights, made rights transferable, or established caps and constraints between systems. Although the case study applications signalled that water governance issues in most of the countries had an understanding of the links between groundwater and surface water, Asian countries in particular seemed to have a more documented hydrological system, while African countries lacked such information. Another broad issue of concern is the fact that externalities and resource constraints are not that well understood across most of the countries, and no countries are enforcing constraints through the presence of a cap. Only a few case study regions (the two areas in Pakistan) signalled that water extraction was monitored and/or enforced, and very few countries showed any progress with developing trade registers, market information and trustworthy systems. However, in Chapter 6, Reardon-Smith and co-authors discussed how the Mekong region is using some satellite technology and river flow monitoring stations to collect information on river flows, but sharing of data is limited between countries, and especially information on water extraction is minimal.

  Although it is clear that formal water markets are not commonly implemented across Africa or Asia, informal water trading is much more common and prevalent. Our case studies in this book highlight that informal markets are widespread and come in diverse forms. In Chapter 4, Zuo and co-authors highlight how informal water trading has been occurring in the Heihe River Basin, Zhangye City, Zhangye City, China since the 1960s, while formal water trading has been adopted only since 2015. Many challenges exist in China's

Table 15.1   *Overview of the WMRA framework in African and Asian countries*

Key fundamental market assessors	Africa			Asia									
	Mozambique	Tanzania	Zimbabwe	Cambodia	China	India	Laos	Myanmar	Nepal	Pakistan (Punjab)	Pakistan (Sindh)	Thailand	Vietnam
Property rights/ institutions:													
Water legislation	✓	✓	✓	✓	✓	✓	✓	✓	✓	✓	✓	✓	✓
Unbundled rights	✓	✓	✓	×	✓	×	×	×	✓	×	×	×	×
Rights transferable	×	×	×	×	✓	×	×	×	×	×	×	×	✓
Rights enforceable	×	×	✓	×	✓	✓	×	×	×	✓	✓	×	✓
Constraints between connected systems	×	×	×	×	✓	×	×	×	✓	×	×	×	×
Hydrology:													
Documented hydrology system	×	×	×	✓	✓	×	✓	✓	✓	×	×	✓	✓
Understanding of connected systems	✓	✓	✓	✓	✓	×	✓	✓	✓	✓	✓	✓	✓
Future impacts modelled	×	×	✓	✓	✓	×	✓	×	✓	×	×	✓	✓
Trade impacts understood	✓	✓	✓	×	×	×	×	×	×	×	×	×	×
Resource constraints understood	×	×	×	×	✓	×	×	×	✓	✓	✓	×	×
Resource constraints enforced (e.g., cap existence)	×	×	×	×	×	×	×	×	×	×	×	×	×
Externalities/ governance:													

Key fundamental market assessors	Africa			Asia									
	Mozambique	Tanzania	Zimbabwe	Cambo-dia	China	India	Laos	Myanmar	Nepal	Pakistan (Punjab)	Pakistan (Sindh)	Thailand	Vietnam
Strong governance impartiality	x	x	x	x	□	x	x	x	✓	x	x	x	x
Existence of externalities understood	x	x	x	x	x	x	x	x	x	x	x	x	x
Water use monitored	✓	x	✓	✓	x	✓	x	✓	x	x	x	✓	✓
Water use enforced	✓	x	x	x	x	✓	x	x	x	✓	✓	x	x
System type:													
Suitability of water sources for trade	x	x	✓	✓	✓	✓	✓	✓	✓	✓	✓	✓	✓
Transfer infrastructure availability/suitability	x	x	✓	x	✓	✓	x	x	✓	x	x	x	x
Regulation trade requirements	x	x	✓	x	x	x	x	x	x	x	x	x	x
Adjustment:													
Gains from trade	x	x	x	x	✓	✓	x	x	x	✓	x	x	x
Political acceptability of trade	x	x	x	x	✓	✓	x	x	x	✓	✓	x	x
Entitlement registers and accounting:													
Trustworthy systems	x	x	✓	x	✓	x	x	x	x	x	x	x	x
Trade and market information	x	x	x	x	x	x	x	x	x	x	x	x	x
Trade step reached:	1	1	1–2	1	1–2	1	1	1	1	1	1	1	1

Note: x indicates further reform required for that issue in the particular regional example; ✓ indicates that there is good evidence supporting that particular part of the assessment; while a smaller ✓ indicates that there is positive but still limited evidence, and thus room for improvement.

formal water market, and it seems that effective monitoring and enforcement of water use are the most urgent issues. Projected future urbanisation and rural land consolidation will promote progress towards water markets in China.

In Chapter 5, Lountain and co-authors explore groundwater markets in West Bengal, India, and highlight the informal tube-well water trade from the 1970s onwards, which is somewhat similar to the situation described in the Indus Basin Irrigation system in Pakistan in Chapter 8. It was also suggested in the Indian case study that water property rights are hampered by rights to complementary resources (for example, diesel, electricity), and that the WMRA framework does not fully account for these interrelationships between these resources. In the case of Pakistan, the functioning of informal groundwater markets was recently constrained by an energy crisis and lowering groundwater tables in many irrigated areas. Nepal is also seen to have ongoing informal water market trading in Chapter 7, but the lack of licensing of water in many areas will limit any formal trading. Johnson and co-authors argue that for Nepal to progress further, it must define water entitlement and allocation arrangements, and have monitoring and enforcement provisions.

Although it is not foreseen that formal water markets can be established any time soon in Zimbabwe, Tanzania and Mozambique, in Chapter 3 Pittock and co-authors suggest that there is a greater potential for informal water markets in Africa. Part of these informal water markets consist of opportunities where irrigators could be encouraged to improve irrigation management and save water, and informally trade this back to mining, hydropower or urban sectors. A growing mining sector in Zimbabwe may also push developments to formal water markets.

15.2.2   Case Studies in Europe, America and Oceania

Table 15.2 includes the case study countries from the Americas, Europe and Oceania and Asia, and illustrates that only six regions in the world have reached Step 3 of the WMRA framework, where there is continual monitoring and adaptation and innovation of established water market trade. The summary in Table 15.2 highlights the diversity that can exist even within one country; for example, the five regions profiled in the United States (US) have all reached very different stages. The Diamond Valley is somewhere between Steps 1 and 2, while Idaho, Montana, Oregon and Washington are described by Gilson and Garrick in Chapter 13 as being at Steps 2 to 3.

The key fundamental water market characteristics that are missing in many of our case study areas in France, Italy, Spain, the United Kingdom (UK), Chile and Diamond Valley in the US include: the unbundling of rights; transferable rights; understanding trade impacts; the existence of a cap (albeit England and Chile have made some progress in this area); the monitoring

Table 15.2   Overview of the WMRA framework in European, American and Oceanian countries

Key fundamental market assessors	Europe					Sth America	Nth America					Oceania			
	France (Poitevin)	France (Neste)	Italy	Spain	UK	Chile	US (Diamond Valley)	US (Idaho)	US (Montana)	US (Oregon)	US (Washington)	Australia (Northern MDB)	Australia (Southern MDB)	Australia (Tasmania)	NZ (Canterbury)
Property rights/ institutions:															
Water legislation	✓	✓	✓	✓	✓	✓	✓	✓	✓	✓	✓	✓	✓	✓	✓
Unbundled rights	x	x	✓	x	x	x	x	✓	✓	✓	✓	✓	✓	✓	✓
Rights transferable	x	x	x	✓	✓	✓	x	✓	✓	✓	✓	x	✓	✓	✓
Rights enforceable	✓	✓	✓	x	✓	✓	✓	✓	✓	✓	✓	✓	✓	✓	✓
Constraints between connected systems	✓	x	x	x	x	x	✓	✓	✓	✓	✓	✓	✓		
Hydrology:															
Documented hydrology system	✓	✓	✓	✓	✓	✓	✓	✓	✓	✓	✓	✓	✓	✓	✓
Understanding of connected systems	✓	✓	✓	✓	✓	✓	✓	✓	✓	✓	✓	✓	✓	✓	✓
Future impacts modelled	✓	✓	✓	✓	✓	✓	✓	✓	✓	✓	✓	✓	✓	✓	✓
Trade impacts understood	x	x	x	✓	✓	x	✓	x	x	✓	x	x	✓	✓	x

Key fundamental market assessors	Europe					Sth America	Nth America					Oceania			
	France (Poitevin)	France (Neste)	Italy	Spain	UK	Chile	US (Diamond Valley)	US (Idaho)	US (Montana)	US (Oregon)	US (Washington)	Australia (Northern MDB)	Australia (Southern MDB)	Australia (Tasmania)	NZ (Canterbury)
Resource constraints understood	✓	✓	✓	×	✓	✓	✓	✓	✓	✓	✓	✓	✓	✓	×
Resource constraints enforced (e.g., cap existence)	×	✓	×	×	✓	✓	×	✓	✓	✓	✓	×	✓	✓	✓
Externalities/ governance:															
Strong governance impartiality	✓	✓	✓	×	✓	✓	✓	✓	✓	✓	✓	×	✓	✓	✓
Existence of externalities understood	✓	✓	✓	✓	✓	×	✓	×	×	✓	×	×	✓	✓	×
Water use monitored	✓	✓	×	✓	×	✓	×	✓	✓	✓	✓	×	✓	✓	✓
Water use enforced	✓	✓	×	✓	×	✓	×	✓	✓	✓	✓	×	✓	✓	✓
System type:															
Suitability of water sources for trade	✓	✓	✓	✓	✓	✓	✓	✓	✓	✓	✓	×	✓	✓	✓
Transfer infrastructure availability/ suitability	✓	✓	✓	✓	✓	✓	✓	✓	✓	✓	✓	✓	✓	✓	✓
Regulation trade requirements	×	✓	×	✓	✓	✓	×	✓	✓	✓	✓	×	✓	✓	×

Key fundamental market assessors	Europe					Sth America	Nth America					Oceania			
	France (Poitevin)	France (Neste)	Italy	Spain	UK	Chile	US (Diamond Valley)	US (Idaho)	US (Montana)	US (Oregon)	US (Washington)	Australia (Northern MDB)	Australia (Southern MDB)	Australia (Tasmania)	NZ (Canterbury)
Adjustment:															
Gains from trade	x	✓	x	✓	✓	✓	✓	✓	✓	✓	✓	x	✓	✓	✓
Trade acceptability	x	x	x	x	✓	✓	✓	✓	✓	✓	✓	x	✓	✓	x
Entitlement registers and accounting:															
Trustworthy systems	✓	✓	x	x	✓	x	x	✓	✓	✓	✓	x	✓	✓	x
Trade and market information	x	x	x	x	x	x	x	✓	✓	✓	✓	✓	✓	✓	x
Trade step reached:	1	1	1	2	2	2	1–2	2–3				2	2–3	3	2

Note: x indicates further reform required for that issue in the particular regional example; ✓ indicates that there is good evidence supporting that particular part of the assessment; while a smaller ✓ indicates that there is positive but still limited evidence, and thus room for improvement.

Sources: Case studies in this book, Wheeler et al. (2017) and Wheeler and Garrick (2020).

and enforcement of water extraction; trustworthy water registers and trade and market information. Many of these factors are essential to the successful operation of water markets.

England has developed water legislation to enable and encourage water trading, and has environmental protection in place (such as hands-off environmental flows). Caps and enforcement are in place, and trade has been growing in predominance since 2018, with many informal water trades having occurred, as detailed by Bark and Smith in Chapter 11. Although Chile introduced legislation to facilitate markets in 1981, as a response to water scarcity issues, Donoso and co-authors in Chapter 12 argue that market adoption is patchy, and highest in areas of water scarcity. Many issues in Chile constrain water markets, such as: ancient water rights not being monitored or measured; return flows; and groundwater substitutability. Pérez-Blanco in Chapter 10 provides an assessment of the Po River Basin District in Northern Italy, and although it was our only case study in Europe where water rights have been unbundled from land, further water market advancement was currently blocked by legislation because of concerns about third-party impacts.

France (Chapter 9) had some of the strongest basic fundamental water market requirements, such as strong institutions and hydrological information, but is still only at Step One of the framework. In particular, the areas of Poitou Marsh Basin and the Neste system in France have significant institutional development established and hydrological information available, along with a cap on water extractions in place in the Neste system, but enforcement is found to be lacking. Informal water trades and swapping are occurring, but further property right reform such as unbundling and increased social acceptance would be needed before further development of the water market.

Gilson and Garrick in Chapter 13 highlight how local institutional capacity is uneven both within and across state boundaries in the United States, and they argue that enabling conditions must be pursued and coordinated at basin, state, and local levels. Although the US is predominantly at Steps 2 to 3 of the trading framework for many states, there are still reforms to be coordinated, and trade impacts and externalities to be understood. The case study of Canterbury in New Zealand is also illustrative of these factors, and in addition highlighted that the cultural importance of water systems must play more of a role in market issues. The New Zealand case study by Talbot-Jones and Grafton in Chapter 14 highlighted that the WMRA framework does not currently properly account for the social and cultural acceptability of trade, and not including these categories specifically could result in some analysts underestimating their importance.

Indeed, the most sophisticated water market in the world, the Murray–Darling Basin (MDB) in Australia, also illustrates the significant disparity that can exist within even one region. In their journal article 'A tale of two water markets', Wheeler and Garrick (2020) illustrate that the northern Murray–Darling Basin water markets are significantly less developed and at a far more nascent stage (Step 2) than the southern Murray–Darling Basin water markets (Step 3). This is of significant interest, especially considering that Australia is often held up to be the world's leading water management standard. Wheeler and Garrick (2020) provide a number of key insights into why water markets have been much more successful in the southern MDB as compared to the northern MDB:

1. Far greater hydrological connectivity (and public dam storage capacity) in the southern than the northern MDB.
2. Far greater amount of unregulated water entitlements in the northern versus southern MDB.
3. Far greater reliance on groundwater as an irrigation source in the northern than southern MDB, plus greater use of on-farm irrigation storage from flood harvesting.
4. Much higher water usage charges paid in the southern versus the northern MDB.
5. Far more irrigators in the southern than northern MDB.
6. Lower average irrigated area per business in the southern than the northern MDB.
7. Higher monitoring of water extractions in the southern MDB versus northern MDB.
8. Far larger water use homogeneity in the northern (mainly cotton industry) than the southern MDB (cereals/rice, pasture and fruit/nut/vegetables).

Wheeler and Garrick (2020) use the northern Basin example as one that highlights the potential societal cost of putting water market and trade in place where there are not strong property rights and institutions in place. For example, in the north, floodplain harvesting and groundwater extraction are very poorly monitored and enforced, and also there is not strong governance impartiality (as has been shown by the success of irrigator lobbying in changing water extraction rules) (AAS, 2019). Allowing additional trade of unregulated water entitlements may then have detrimental impacts on downstream irrigators or the environment, and should not be allowed until stronger institutions, an effective cap, monitoring and measurement of all extractions are in place (Wheeler and Garrick, 2020). Wheeler et al. (2020a) also emphasised the consequences for water extraction within a region where substantial subsidies for irrigation infrastructure exist. It was shown that irrigators who received irrigation infrastructure subsidies in the southern Murray–Darling Basin actually increased their water extraction as a result (confirming the rebound effect of subsidies on water extraction often discussed in the literature, e.g.,

Gómez and Pérez-Blanco, 2014), as compared to all other irrigators (which included irrigators who privately financed their upgraded irrigation infrastructure, or irrigators using water markets or irrigators who did not perform these adaptation actions) (Wheeler et al., 2020a). As the groundwater case study in India showed, electricity subsidies for pumping groundwater led to increased water extraction there as well, highlighting the fact that subsidies often distort decision-making and can worsen water scarcity issues.

15.3 DISCUSSION

15.3.1 What are the Benefits and What are the Market Failures of Water Markets?

As a reminder from Chapter 1, economic studies have highlighted that there are three distinct forms of economic efficiency associated with water markets that make them important as a tool for dealing with water scarcity issues in a country: (1) allocative efficiency: improving water resource short-term decision-making, reflecting seasonal conditions, is facilitated by short-term water trade; (2) dynamic efficiency: improving or facilitating water resource structural or long-term decision-making, reflecting new investment opportunities, regulatory shifts in access arrangements or personal strategic choices, can be achieved through long-term water trade; and (3) productive efficiency: increasing the flexibility of water prices offers incentives for the efficient use of water resources, as either an investment or input for productive outcomes (Grafton and Wheeler, 2018). In particular, when water markets are designed properly and have strong institutions supporting and governing water extraction and consumption, the individual benefits of a water market include: it allows water to be traded to its highest value use (including urban and environmental); involves only willing buyers and sellers, and hence provides some security tenure over transactions; supports long-term farm development; provides a risk management strategy for farmers; provides flexibility and additional income stream for annual growers in times of high water scarcity and a source of much-needed water for permanent growers; reduces probability of bankruptcy during drought; allows purchase for environmental (or cultural) benefits and the same rights as irrigation holders; can free up capital for farmers to use elsewhere; increases water entitlement value and asset values of irrigators; movement of water can have positive environmental impact; and allows non-landholders to enter the market, who often develop new innovative risk products, and their increased demand in the market can increase water values for existing users (Dinar et al., 1997; Easter and Huang, 2014; Easter et al., 1998; Griffin, 2006; Griffin et al., 2013; Wheeler, 2014; Wheeler et al., 2014a, 2014b, 2017, 2020a).

On the other hand, the costs of water markets include the existence of any market failures (see Quiggin, 2019 for a full discussion on market failure). The major types of market failures include:

1. Imperfect competition. This will occur if output markets are not contestable but nevertheless characterised by monopoly, oligopoly, bilateral monopoly or some other market imperfection.
2. Externalities. This will occur when property rights are not clearly defined, and so costs and/or benefits observe spillovers to others. In this case, discrepancies between private and social benefits and costs will be observed, and the resource allocation generated by markets will not be efficient because market prices do not reflect the 'full' or social costs involved.
3. Information asymmetry. This describes the situation where one party of a transaction has better information than the other. In this case, the information-rich agent can act in their own interests at the cost of the information-poor. Two typical problems are adverse selection and moral hazard (Quiggin, 2019).

All water markets are subject to market failure. Indeed, the most sophisticated and adopted water market in the world, in Australia, has been shown to exhibit numerous water market failures (Seidl et al., 2020b; Wheeler and Garrick, 2020). In particular, imperfect competition exists in the Murray–Darling Basin, especially with regards to the northern Basin, inter-valley trade issues and unregulated water broker behaviour. Negative externalities are also clearly present, mainly because of the lack of clear property rights and institutional rules. Information asymmetry is also clearly present in water markets, in relation to data and information on prices, water registers and weather, insider trading issues of working groups and water brokers, to name but a few issues.

In terms of equity issues, given that water markets can result in increases/decreases in water prices, some associate this pecuniary externality as a negative/cost of water markets. However, when markets are complete, pecuniary externalities do not matter; but they do matter when markets are subject to market failure, and hence the welfare effects of a price movement may not offset each other. Issues with resource constraints, capacity and contract issues matter here, and so does the initial distribution of property rights. Again, for the case of the Murray–Darling Basin, which is at Step 3 of the WMRA framework, the continual need for adaptation, adjustment and reassessment is critically important (Wheeler et al., 2017). The initial distribution of property rights will also matter for water resources that have been identified as having the potential for increased extraction (for example, the water resources in the Northern Territory, Australia, provide one such case study), and proper consideration of how to create new licences and their governance rules will be critical to avoid water market failures.

15.3.2 Key Lessons Learned from the Case Studies

The application of the WMRA framework to 20 regions across six continents in the world in this book has highlighted its usefulness as a test to see how 'ready' a country may be for the implementation of formal water markets. Overall, formal water markets are not possible for many countries across the world to implement, as they have not achieved the first steps of establishing property rights, caps on water extraction, strong independent water institutions and governance impartiality; and if formal trade is introduced, this may result in a situation of increased water extraction and worsening water scarcity issues. For example, if water trade and exchange is permitted and there is not a cap set on water extraction – or if it is set above current extraction levels – it is possible that trade will result in increased water extraction as formal trade occurs and activates former unused licences. The availability of 'new' water could also lead to unanticipated increases in irrigated area and an increased demand for water (Grafton et al., 2018). Subsidies on electricity use, irrigation infrastructure and improper pricing of regional operations and maintenance also leads to negative externalities and increased water extraction in many countries around the world.

Hence, it is critical that there is a correct sequencing of water reforms before formal water trade can occur (Young, 2019). Some of the serious outstanding issues that remain for many countries to deal with include the following.

First, establishing sustainable (and adaptable) water extraction caps. The importance of establishing sustainable water extraction (groundwater and surface water) caps cannot be over-emphasised. Even for countries which have done so, such as the Murray–Darling Basin in Australia, the incomplete knowledge of the interconnectedness and substitutability between difference water resources means that a cap on surface water extraction but less stringent controls on groundwater may mean that official cap figures can be exceeded (Wheeler et al., 2020a, 2020b). Indeed, the very nature of water scarcity factors in many countries means that water extraction caps require greater definition. England, for example, has introduced a range of environmental protections and the concept of 'hands-off' environmental flows. Young (2019) argued that the water sharing pool needs to be defined on a reach by reach (and groundwater by groundwater) basis, with 'hands-off flow', or conveyance water, built into a complete water sharing system. Such a measure would incorporate connectivity issues, and once this threshold is reached, no other allocations by users would be available. The next level of sharing would go from high priority to general, to low priority, which the environment has a range of different shares in. For example, the environment may have a lower amount of high-priority shares in such a system, and correspondingly have a greater proportion of low-priority shares (AAS, 2019; Young, 2019). Wheeler et al. (2020b) and Schwabe et al. (2020) also highlighted the potential relationship between increased water market trade out of a region, and increased groundwater extraction.

Second, water accounting. Basic hydrological information such as sound measurement of all inflows, water consumption, recoverable return flows, and return flows to sinks, need to be included in a water accounting framework. In addition, the development of markets needs standardised and transparent methodology for measuring licence values and providing information for trade (Seidl et al., 2020a, 2020b; Wheeler et al., 2020b).

Third, measuring, monitoring and enforcing extractions. It is clear from the case studies that most countries do a poor job in monitoring and measuring water extraction, let alone enforcing water extraction. The continual development of satellite and thermal technology in measuring water extraction and consumption will be the most cost-effective measure for countries to adopt in the future, and is a space to watch. All water extraction must be transparent, fully audited and underpinned by sound measurement of all inflows, water consumption, recoverable return flows, and flows to sinks; and extraction must be enforced, with appropriate penalties put in place for illegal behaviour (Wheeler et al., 2020b).

Fourth, cultural values. The cultural values of rivers of indigenous owners in various countries have traditionally been ignored, which is especially the situation for many developed countries in their initial distribution of property rights. There remain many regions where there is the potential for indigenous owners to be given a greater share of water ownership when creating new water licences. For countries or regions developing formal water markets, such cultural rights must be considered in the initial design of bestowing water rights (the Northern Territory provides an Australian example for this, before they create more water rights), while for those regions which have developed formal water markets, how to address the equity issues of existing rights distribution must be considered. Talbot-Jones and Grafton discuss this in depth for the New Zealand case study (Chapter 14).

The lessons above apply the most to countries or regions developing water markets, but many elements also apply to regions with the most sophisticated water markets. Water market institutions represent a continual journey of adaptation as external factors change. Indeed, continual work on issues such as transaction costs, transparency, information flows and new products will all help water markets to achieve their true potential (Schwabe et al., 2020; Seidl et al., 2020a, 2020b).

15.4 CONCLUSION

Overall, the WMRA framework, as applied to 20 regions across six continents in the world in this book, has been shown to be a valuable tool in understanding the limitations and the gaps that many regions in the world have before more formal water markets can be developed. Although it is clear that water markets can bring numerous benefits as another demand management tool that can be employed to help deal with water scarcity, it is also clear that they need very careful implementation, and that markets only exist within institutions and structures which allow and govern the transfer of water. If these institutions and structures are corrupted or are missing, then this can result in negative impacts for society from the implementation of water markets. Similarly, the presence of subsidies (in either resource use or for irrigation infrastructure provision) distorts both decision-making and efficient water extraction. The most sophisticated water markets in the world, such as in the Murray–Darling Basin, highlight the need for continual adaptation and adjustment of water markets, as well as this dual case of potential negative impacts from the implementation of markets in regions (such as the Northern Murray–Darling Basin) that do not have strong institutions, caps for all forms of water extraction, monitoring and measurement in place.

As such, although formal trade will not be possible for many countries, considerable informal water market trading exists in various countries, whether it be via informal arrangements or via access to groundwater tube-wells and pumping, and subsidised by electricity use. Increasing informal trade is not only an important step to formal water markets, but it is an important step in itself, given that most countries will probably not be able to progress to formal water markets. Increased promotion of informal markets, and simple 'swapping' of water, should be encouraged to promote economic efficiencies. However, reform must still continue in regard to the removal of subsidies that distort resource use and decision-making, such as subsidised electricity use in India and subsidised irrigation infrastructure in many countries around the world. In addition, more opportunities may also exist for expanding water quality markets.

In the future, the continual evolution and development of satellite and thermal technology in monitoring and measuring both surface and groundwater extractions and consumptive use may provide one way to help further establish caps, hands-off flows, trade and externality impacts and an understanding of resource constraints; however, these factors need to be coupled with enforcement, and unbundled and transferable rights. Further consideration of cultural property rights, the interconnectedness of water resources, human behavioural consequences from implementing markets, and externality and other market failure issues, is also needed.

REFERENCES

AAS (2019). *Investigation of the Causes of Mass Fish Kills in the Menindee Region NSW Over the Summer of 2018/2019*. Australian Academy of Science, Canberra.
Barbier, E. (2019). *The Water Paradox: Overcoming the Global Crisis in Water Management*. Yale University Press.
Dinar, A., Rosegrant, M.W., and Meinzen-Dick, R. (1997). *Water Allocation Mechanisms: Principles and Examples*. Policy Research Working Paper, World Bank, no. 1779. http://elibrary.worldbank.org/doi/abs/10.1596/1813-9450-1779.
Easter, K., and Huang, Q. (eds) (2014). *Water Markets for the 21st Century*. Springer.
Easter, K., Rosegrant, M., and Dinar, A. (1998). *Markets for Water: Potential and Performance*. Natural Resource Management and Policy series. Kluwer Academic Publishers.
Gómez, C., and Pérez-Blanco, C., 2014. Simple myths and basic maths about greening irrigation. *Water Resources Management, 28*, 4035–4044.
Grafton, R. and Wheeler, S.A. (2018). Economics of water recovery in the Murray–Darling Basin, Australia. *Annual Review of Resource Economics, 10*(1), 487–510.
Grafton, R., Williams, J., Perry, C., Molle, F., Ringler, C., et al. (2018). The paradox of irrigation efficiency. *Science, 361*(6404), 748–750.
Griffin, R.C. (2006). *Water Resource Economics: The Analysis of Scarcity, Policies, and Projects*. MIT Press.
Griffin, R.C., Peck D.E., and Maestu J. (2013). Introduction: myths, principles and issues in water trading. In J. Maestu (ed.), *Water Trading and Global Water Scarcity: International Experiences* (pp. 1–14). RFF Press Water Policy Series.
Quiggin, J. (2019). *Economics in Two Lessons: Why Markets Work So Well, and Why They Can Fail So Badly*. Princeton University Press.
Schwabe, K., Nemati, M., Landry, C., and Zimmerman, G. (2020). Water markets in the western United States: trends and opportunities. *Water, 12*, 233.
Seidl, C., Wheeler, S.A., and Zuo, A. (2020a). High turbidity: water valuation and accounting in the Murray–Darling Basin. *Agricultural Water Management, 230*, p105929.
Seidl, C., Wheeler, S.A., and Zuo, A. (2020b). Treating water markets like stock markets: key water market reform lessons in the Murray–Darling Basin. *Journal of Hydrology, 581*, 124399.
Wheeler, S.A. (2014). Insights, lessons and benefits from improved regional water security and integration in Australia. *Water Resources and Economics, 8*, 57–78.
Wheeler, S.A., and Garrick, D.E. (2020). A tale of two water markets in Australia: lessons for understanding participation in formal water markets. *Oxford Review of Economic Policy, 36*(1), 132–153.
Wheeler, S.A., Carmody, E., Grafton, R., Kingsford, R., and Zuo, A. (2020a). The rebound effect on water extraction from subsidising irrigation infrastructure in Australia. *Resources, Conservation and Recycling, 159*, 1–17.
Wheeler, S.A., Loch, A., Crase, L., Young, M., and Grafton, R. (2017). Developing a water market readiness assessment framework. *Journal of Hydrology, 552*, 807–820.

Wheeler, S., Zuo, A., and Bjornlund, H. (2014a). Investigating the delayed on-farm consequences of selling water entitlements in the Murray–Darling Basin. *Agricultural Water Management*, *145*, 72–82.

Wheeler, S., Zuo, A., and Hughes, N. (2014b). The impact of water ownership and water market trade strategy on Australian irrigators' farm viability. *Agricultural Systems*, *129*, 81–92.

Wheeler, S., Zuo, Z., and Kandulu, J. (2020b). What water are we really pumping? The nature and extent of surface and groundwater substitutability in Australia and implications for water management policies. *Applied Economic Perspectives and Policy*. https://doi.org/10.1002/aepp.13082.

Young, M. (2019). *Sharing Water: The Role of Robust Water-Sharing Arrangements in Integrated Water Resources Management*. Perspectives paper by Global Water Partnership. https://www.gwp.org/globalassets/global/toolbox/publications/perspective -papers/gwp-sharing-water.pdf.

Index